Library of Congress Cataloging-in-Publication Data

MacCluer, C. R.
    Industrial Mathematics : modeling in industry, science, and government / Charles R. MacCluer.
        p. cm.
    Includes bibliographical references (p.
    ISBN: 0-13-949199-6
    1. Mathematical models.   I. Title.
QA401.M33 2000
511' .8--dc21   99-27073
                                                        CIP

Acquisitions Editor: *George Lobell*
Editorial/Production Supervision: *Rick DeLorenzo/Bayani Mendoza de Leon*
Editor-in-Chief: *Jerome Grant*
Assistant Vice President of Production and Manufacturing: *David W. Riccardi*
Senior Managing Editor: *Linda Mihatov Behrens*
Executive Managing Editor: *Kathleen Schiaparelli*
Manufacturing Buyer: *Alan Fischer*
Manufacturing Manager: *Trudy Pisciotti*
Marketing Manager: *Melody Marcus*
Marketing Assistant: *Amy Lysik*
Art Director: *Jayne Conte*
Assistant to the Art Director: *Bruce Kenselaar*
Editorial Assistant: *Gale Epps*
Cover photo: From Murphy/Jahn Architects, Inc., "The Master Architect Series, Murphy/Jahn, Selected and Current Works," copyright 1995 Images Publishing Group Pty Ltd., render/photograph by Michael Budilovsky.

Printed in the United States of America

10 9 8 7 6 5 4 3 2

ISBN 0-13-949199-6

Prentice-Hall International (UK) Limited, *London*
Prentice-Hall of Australia Pty. Limited, *Sydney*
Prentice-Hall Canada Inc., *Toronto*
Prentice-Hall Hispanoamericana, S.A., *Mexico*
Prentice-Hall of India Private Limited, *New Delhi*
Prentice-Hall of Japan, Inc., *Tokyo*
Prentice-Hall (*Singapore*) Pte. Ltd.
Editora Prentice-Hall do Brasil, Ltda., *Rio de Janeiro*

Industrial Mathematics

Modeling in Industry, Science, and Government

Charles R. MacCluer
Michigan State University

PRENTICE HALL, Upper Saddle River, New Jersey 07458

*Dedicated
to
my mother and father, Colen and Theron.*

Contents

Preface

About This Book

Mathematics is unreasonably effective in resolving seemingly intractable problems. The process proceeds in three steps: model the external world problem as a mathematical problem, solve the mathematical problem, then interpret the results. A mathematician in government or industry will be involved at all three steps.

This book is for students about to enter the workforce. They may be well grounded in the fundamentals of mathematics but not in its practice. Although changing of late through the efforts of COMAP, SIAM, and NSA, the graduating student has little experience in modeling or in the particular extensions of mathematics useful in industrial problems. They may know power series but not the z-transform, orthogonal matrices but not factor analysis, Laplace transforms but not Bode plots. Most certainly they will have no experience with problems incorporating the unit \$. Mathematicians in industry must be able to see their work from an economic viewpoint. They must also be able to communicate with engineers using their common dialect, the dialect of this book.

Each chapter begins with a brief review of some relevant mathematics which may require further elaboration by the instructor. Then the industrial extension of this same material is introduced via typical applications. The routines which occur in the flow of text are not merely enrichment but instead are an integral part of the text itself. One central thrust of this book is to demonstrate the power of interweaving analytic with computing methods during problem solving.

Many exercises require the student to experiment with, or to modify, the MATLAB routines provided. Tedious retyping of the routines is unnecessary since all routines will be available at our anonymous ftp site *math.msu.edu* down the path *pub/maccluer/mm*. Other exercises ask the student to generate code themselves. A certain number of exercises are in fact projects, requiring data collection, experimentation, or consultation with industrial experts.

This book is aimed at the senior undergraduate or Master's student of mathematics, engineering, or science. The writing style is by design sparse and brief.

To the Instructor

Let me tell you how I use this book. I feel it is crucial that students obtain experience in group project development. The nearly universal opening gambit of job interviewers is to ask student applicants to describe their group project experience.[1] To provide this experience I require three projects to be completed during the course — the first two done by groups of size two or three, the last solo. The "deliverable" from each project is a formal technical report described in Chapter 15.

The projects are chosen with my advice and consent. Often, student groups will propose their own project or a variation on a project from the book. I have had to maintain an open door policy for frequent consultations with the groups as they develop their projects. I return flawed reports and ask that they be resubmitted. But in the end I have been *astonished* by the quality of many of the finished reports.

Once reports have been completed, I ask each group to select a member to deliver their report to the class as a 12-minute overhead slide presentation (after reading the *do's* and *don'ts* of such talks in Chapter 15). This experience is transformational for the student.

I strongly believe in weekly homework — lots of it. I encourage my students to work in study groups but require write-ups to be done individually. Above all, *each student must do simulation and numerical experimentation individually.* The worst case is for one member of a study group to do all the computer work for the group. Simulation homework is in the form of a memo, plus source code, plus data. I point out that source code is more individual than fingerprints and have been able to head off this problem from the start. A major objective of my course is for *each* student to develop a symbiosis with their silicon-based helpmate.

Students lack experience in taking "first cuts," at biting off pieces of a problem. I constantly urge them to cut the problem down to solvable size, estimate, do a special case. I ask, "What could you have on my desk before 5:00?" I encourage running off to the computer. Insisting on elegant analytic solutions is not cost-effective and is not in the spirit of this course (or of industry).

---

[1] The second question is usually about how they handled group members who did not carry their load.

Ideally, this course would be followed by an *Industrial Projects* tutorial course, where local industrial representatives would propose problems, then serve (with a faculty member) as liaison to a student project group as they develop a project during the term. At term's end the group would present a formal technical report to both the participating faculty group and to the industrial group. This is patterned after a very successful summer program developed by H. T. Banks and H. T. Tran of North Carolina State University in cooperation with the National Security Agency.

Chapter Interdependence

This book is in large part a collection of independent topics, a survey of the mathematics essential to an industrial mathematician. The only iron-clad dependence is Chapter 2 on Chapter 1. Chapter 10 depends on some basic notions from a sophomore differential equations course reviewed in Chapter 9. Inner product notation is introduced in Chapter 6 and used in Chapters 11–14, but could be introduced as needed. The numerical methods described in Chapters 12–14 could be taught without covering Chapters 9 and 11 by relying only on student experiences from previous courses.

The Symbol *

A single asterisk (*) beside an exercise indicates that the exercise requires some result or technique from an advanced senior or first-year graduate course, or that it may be a bit more difficult. A double asterisk signals that the exercise is quite difficult. On several occasions an asterisk is used to indicate a section or a proof at an advanced level.

Close the Loop

Please let me know about your successes and failures with this book. Above all, tell me about successful projects not suggested in the book. I will post your projects on my Web site:

http://www.math.msu.edu/ maccluer/index.html

Acknowledgments

This book grew from conversations with George Lobell, mathematics editor of Prentice Hall, but many others contributed to its final form. I am grateful for expert advice from R. Aliprantis, T. V. Atkinson, M. C. Belcher, R. V. Erickson, D. Gilliland, E. D. Goodman, Jon Hall, Frank Hatfield, T. J. Hinds, F. C. Hoppensteadt, R. J. LaLonde, D. Manderscheid, L. V. Manderscheid, Mark S. McCormick, R. Narasimhan, G. L. Park, R. E. Phillips, Jacob Plotkin, P. A. Ruben, W. E. Saul, V. Sisiopiku, Lee Sonneborn, V. P. Sreedharan, G. C. Stockman, T. Volkening, and John Weng.

I am also indebted to the four reviewers who provided suggestions for improving the material covered in this book. They are Lester F. Caudill, Jr., University of Richmond; Gary Ganser, West Virginia University; Reinhard Illner, University of Victoria; Anne Morlet, Argonne National Laboratory; and Alan Struthers, Michigan Technological University.

Clark J. Radcliffe was an especially good resource for engineering matters.

Special thanks to Karen Holt, Technical Writing Consultant of the Naval Undersea Warfare Center Division, for assistance with Chapter 15.

Many undergraduate students helped shape this book; I thank M. Alexander, M. S. Bakker, Richelle Brown, C. Crews, R. L. Ennis, J. M. Foster, P. Franckowski, M. C. Hilbert, Matt Johnson, D. Karaaslanli, J. A. Leikert Jr., K. S. Little, Shana Ostwald, D. M. Padilha, Faisal Shakeel, J. K. R. Shirley, J. R. Shoskes, R. D. Smith, K. A. Tillman, and B. R. Tyszka.

But much credit belongs to a talented group of graduate students: E. Andries, Reena Chakraborty, K. L. Eastman, M. B. Hall, E-J Kim, J-T Kim, F. Kivanc, S. A. Knowles, J. R. Lortie, T-W Park, and R. L. Rapoport. Good luck to you in your industrial careers.

C. R. MacCluer

Chapter 1

Statistical Reasoning

Individual behavior may be erratic, but aggregate behavior is often quite predictable.

Statistics is second only to differential equations in the power to model the world about us. Statistics enables us to predict the behavior on average of systems subject to random disturbances. This chapter begins with a review of random variables and their cumulative distributions. It is followed by four sections defining the four most useful distributions: the uniform, Gaussian, binomial, and Poisson. The power of statistical reasoning will be apparent in the models constructed in each section. The concluding section, §1.6, is a brief introduction to the *Taguchi off-line quality control* approach that has revolutionized Japanese industry.

§1.1 Random Variables

An experiment is performed repeatedly, say a person is selected from a large population. A measurement X is taken, say the person's height. After all experimentation is done, we have statistical data about the population — the *random variable X* gives rise to a *cumulative probability distribution function $F(x)$*:

$$F(x) = \text{probability that } X \leq x \qquad (1.1)$$

such as shown in Figure 1.1. We assume that the populations are large and that measurements are made with infinite precision so that we are justified in replacing bins of measurement intervals, each with its own incidence of occurrence, with a continuum of probabilities as shown in Figure 1.1. Thus

$$\text{prob}(a \leq X \leq b) = \int_a^b dF(x) = F(b) - F(a^-) \qquad (1.2)$$

1

and more generally,

$$\text{prob}(X \in S) = \int_S dF(x). \qquad (1.3)$$

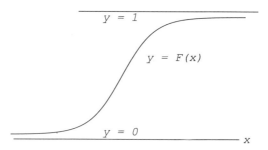

Figure 1.1. Continuous cumulative distribution function.

Example. What is the life span T (in hours) of a typical 60 W hotel hallway light bulb? The associated cumulative distribution function $F(t)$ is the percentage of the bulbs that have burned out at or before time t. We can with confidence anticipate that $F(t)$ is 0 before $t = 0$, rising slowly at first, then more quickly as we near and pass the typical life span (1000 hours), thereafter leveling off to approach 1 asymptotically from below. Our intuition is confirmed by the following mortality data supplied by General Electric:

hrs	400	500	600	700	800	900	1000	1100	1200	1300	1400
fail	0%	2%	5%	10%	20%	30%	50%	70%	80%	90%	95%

The $dF(x)$ in (1.3) reflects the intuition that since X takes on some values more often than others, it gives rise to a new measure on intervals: the new length of the interval $[a, b]$ is the probability that the values of X fall into $[a, b]$, thus the equation (1.2). Employing this new measure of intervals, one constructs the (*Stieltjes*) integral as before but now with the infinitesimal $dF(x)$ rather than dx ([Cramér]).

For any function $Y = g(X)$ of the random variable X, say the conversion from British to metric units, the *expected value* of the random variable $Y = g(X)$ is

$$E[Y] = \int_{-\infty}^{\infty} g(x) \, dF(x). \qquad (1.4)$$

In particular,

$$\mu = E[X] = \int_{-\infty}^{\infty} x\, dF(x) \qquad (1.5)$$

is the *mean* or *center of mass* of the random variable X, while

$$\nu = \sigma^2 = E[(X - \mu)^2] = \int_{-\infty}^{\infty} (x - \mu)^2\, dF(x) = E[X^2] - \mu^2 \quad (1.6)$$

is the *variance* or *moment of inertia about the center of mass*, and the quantity $\sigma = \sqrt{\nu}$ is the *standard deviation* of X. Examples are given below.

If $F(x)$ has no jumps and is in fact differentiable, the derived function

$$f(x) = F'(x) \qquad (1.7)$$

is called the *probability density function*, so that

$$\text{prob}(g(X) \in S) = \int_S g(x)\, dF(x) = \int_S g(x) f(x)\, dx. \qquad (1.8)$$

Example. Due to a fundamental problem of measurement, the location and momentum of a very small particle cannot both be predicted with certainty. All that is knowable is its *wave function* ψ and hence a probability density function $f(x) = |\psi|^2$. (Think of crimes making up a crime wave.) This wave function ψ is a time-weighted superposition of eigenfunctions (*stationary states*) of the Hamiltonian operator H, the instrument that observes total energy. For example, the state of lowest energy (*ground state*) of a quantum particle trapped in the potential free interval $[0, 1]$ is $\psi_0 = \sqrt{2} \sin \pi x$ [MacCluer, 1994, p. 134]. Thus the expected (mean) position of a particle in this state is

$$\int_{-\infty}^{\infty} x\, dF(x) = \int_{-\infty}^{\infty} x f(x)\, dx = 2 \int_0^1 x \sin^2 \pi x\, dx = \cdots = 0.5.$$

You cannot hope to dodge some working knowledge of quantum mechanics since it has become routine in industrial applications as mundane as measuring NO_x in engine emissions. More on this will appear in Chapter 11.

At the other extreme, a cumulative distribution F may be *discrete*, consisting of steps of height m_i at $x = x_i$ so that

$$\text{prob}(g(X) \in S) = \int_S g(x)\, dF(x) = \sum_{x_i \in S} g(x_i) m_i, \qquad (1.9)$$

as shown in Figure 1.2.

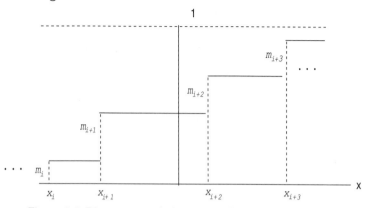

Figure 1.2. Discrete cumulative probability distribution function.

Example. Write out π in base 2. It is true but difficult that 0 and 1 occur with equal frequency in this expansion. Thus the cumulative distribution $F(x)$ of the random variable X that returns the entry in a random digit of the expansion is

$$F(x) = \begin{cases} 0 & \text{if } x < 0 \\ 1/2 & \text{if } 0 \le x < 1 \\ 1 & \text{if } 1 \le x \end{cases}$$

with mass $1/2$ at both $x = 0, 1$. (This is of course also the distribution belonging to the experiment of flipping a fair coin with heads $X = 1$ and tails $X = 0$.) Thus for any function $Y = g(X)$ of X,

$$E[g(X)] = \int_{-\infty}^{\infty} g(x)\, dF(x) = g(0)(1/2) + g(1)(1/2).$$

In particular, the mean $\mu = 0 \cdot (1/2) + 1 \cdot (1/2) = 1/2$ and the variance $\nu = (0 - 1/2)^2 (1/2) + (1 - 1/2)^2 (1/2) = 1/4$, giving a standard deviation of $\sigma = 1/2$.

Two random variables X and Y are *independent* of one another if information about the values of one does not prejudice the values possible for the other, i.e.,

$$\text{prob}(X \in S \text{ and } Y \in T) = \text{prob}(X \in S) \cdot \text{prob}(Y \in T),$$

so

$$E[f(X) \cdot g(Y)] = E[f(X)] \cdot E[g(Y)]. \tag{1.10}$$

Example. Two fair dice rolled simultaneously yield top face values X and Y that are independent. In contrast the top and bottom face values of one dice are not independent.

§1.2 Uniform Distributions

Suppose that any outcome in the interval $[a, b]$ is equally likely and no other outcome is possible, i.e., the random variable X has cumulative distribution function

$$F(x) = \begin{cases} 0 & \text{if } x < a \\ \dfrac{x-a}{b-a} & \text{if } a \le x \le b \\ 1 & \text{if } b < x. \end{cases} \qquad (1.11)$$

Then X is said to be *uniformly distributed* on $[a, b]$. Clearly, X has mean

$$\mu = \int_{-\infty}^{\infty} x \, dF(x) = \frac{1}{b-a} \int_a^b x \, dx = \frac{a+b}{2} \qquad (1.12)$$

and variance

$$\nu = \sigma^2 = \frac{1}{b-a} \int_a^b (x - \mu)^2 dx = \frac{(b-a)^2}{12} \qquad (1.13)$$

(Exercise 1.6). In general,

$$E[g(X)] = \frac{1}{b-a} \int_a^b g(x) \, dx. \qquad (1.14)$$

Example. The randomly sampled angle of a turning fan blade is uniformly distributed between 0 and 2π with mean $\mu = \pi$ and standard deviation $\sigma = \pi/\sqrt{3}$.

§1.3 Gaussian Distributions

A random variable X that measures outcomes determined by many and various influences is often *normally* distributed with probability density

$$f(x) = \frac{1}{\sqrt{2\pi}\,\sigma} e^{-(x-\mu)^2/2\sigma^2} \qquad (1.15)$$

with mean μ and standard deviation σ. This is the familiar bell shaped distribution shown in Figure 1.3.

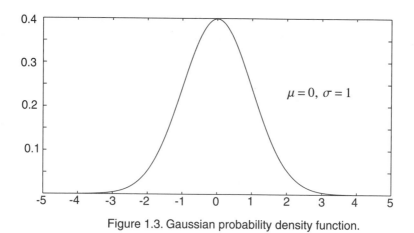

Figure 1.3. Gaussian probability density function.

Example. A certain scholarship aptitude test is designed under the belief that scores should be normally distributed with mean 500 and standard deviation 100. The percentage of students with scores between say 550 and 675 can be deduced by normalizing to the case $\mu = 0$ and $\sigma = 1$ with the change of variables

$$Y = \frac{X - \mu}{\sigma}. \tag{1.16}$$

The score range $550 < X < 675$ becomes $0.5 < Y < 1.75$. Consulting tabular data found in any book on statistics, the integral

$$\text{prob}(0.5 < Y < 1.75) = \frac{1}{\sqrt{2\pi}} \int_{0.5}^{1.75} e^{-y^2/2} \, dy \approx 27\%. \tag{1.17}$$

Alternatively one may write a short routine in MATLAB to approximate the integral of (1.17) after integrating the Taylor series expansion of the integrand term by term. Or most simply, execute this short *Mathematica* script:

```
In[1]:= f[x_]:= Exp[-y*y/2]/Sqrt[2*Pi]
In[2]:= NIntegrate[f[y],{y,0.5,1.75}]
```

With this script we see that a normally distributed outcome X is clustered about its mean as follows:

$$\text{prob}(-\sigma < X - \mu < \sigma) \approx 0.6827,$$
$$\text{prob}(-2\sigma < X - \mu < 2\sigma) \approx 0.9545,$$
$$\text{prob}(-3\sigma < X - \mu < 3\sigma) \approx 0.9973,$$
$$\text{prob}(-4\sigma < X - \mu < 4\sigma) \approx 0.9999.$$

§1.4 The Binomial Distribution

Suppose that there are n devices, each with a probability p of failure during a given time period. What is the probability that exactly k fail during this time period?

Choose a subset of the devices of cardinality k. The probability that all these k devices fail but no others, is $p^k(1 - p)^{n-k}$. Thus multiplying by the number of ways of choosing subsets of size k gives the probability that exactly k fail as

$$p(k) = \binom{n}{k} p^k(1-p)^{n-k}. \tag{1.18}$$

This is the *binomial distribution,* a discrete distribution with cumulative distribution function

$$F(x) = \sum_{k \leq x} \binom{n}{k} p^k(1-p)^{n-k}, \tag{1.19}$$

with mean

$$\mu = pn \tag{1.20}$$

and standard deviation

$$\sigma = \sqrt{np(1-p)} \tag{1.21}$$

(Exercise 1.8). The binomial distribution models well the costs associated with the clumping of a finite number of events when they are all of equal probability of occurrence.

Problem. A campaign staffer knows from experience that only one in every three volunteers called will actually show up to distribute leaflets. How many phone calls must be made to guarantee at least 20 workers with a confidence of 90%?

Solution. The staffer must make n phone calls where the probability that at most 19 show is less than 10%, i.e., where

$$\sum_{k=0}^{19} p(k) = \sum_{k=0}^{19} \binom{n}{k} \left(\frac{1}{3}\right)^k \left(\frac{2}{3}\right)^{n-k} < 0.1. \tag{1.22}$$

Let us test this inequality with MATLAB for $60 \leq n \leq 80$.

```
% Routine 1.1
p = 1/3;                          % one out of three show up
for n=60:80                       % start with 60 calls
  sum = 0;                        % initialize the sum
  pk = (1-p)^n;                   % this is p(0)
  for k=1:19                      % compute p(k) and sum
    pk = (n-k+1)*p*pk/k/(1-p);    % p(k) from p(k-1)
    sum = sum + pk;              % add in the kth contrib.
  end;                           % sum complete for this n
  E(n) = sum;                     % tabulate summations
end;                             % tried up to n = 80 calls
plot(E)                          % graph falls below 0.1?
```

The routine yields a plot showing that $n \approx 75$; in fact, $E(73) = 0.1137$ and $E(74) = 0.0995$, so $n = 74$ calls must be made.

Here is a classic of operations research with wide application in cost/benefit questions in commerce and warfare [Morse and Kimball, p. 31].

The Newsboy Problem. A newsboy sells newspapers outside Grand Central Station. He has on average 100 customers per day. He buys papers for 50 cents each, sells them for 75 cents each, but cannot return unsold papers for a refund. How many papers should he buy?

Binomial Solution. Each of the (say) $n = 5000$ persons passing by the newsboy during the day has the probability $p = 100/n$ of purchasing a paper. Thus the probability that he would encounter exactly k potential customers is

$$p(k) = \binom{n}{k} p^k (1-p)^{n-k}. \qquad (1.23)$$

Suppose that he buys m papers. His profit on days with exactly k potential customers would be $0.75k - 0.5m$ provided that $k \leq m$. On days when $k > m$, his profit is $0.25m$. Distinguishing days by the number k of potential customers yields the expected profit of

$$E[\text{profit}] = \sum_{k=0}^{m} (0.75k - 0.5m) \, p(k) + \sum_{k=m+1}^{n} 0.25m \cdot p(k). \qquad (1.24)$$

It remains to see for which m this expected profit reaches its maximum value. This computation is numerically sensitive since it involves the product of many large by many small numbers. A brute-force calculation is nevertheless successful in MATLAB.

```
% Routine 1.2
n = 5000;                        % number of passersby
p = 100/n;                       % probability of a sale
clear E;                         % clear the variable E
for m=85:105                     % buy m papers
  pk= (1-p)^n;                   % initialize p(0)
  e = -.5*m*pk;                  % loss with no sales
  for k=1:m                      % sum profit for k <= m
    pk = (n-k+1)*pk*p/((1-p)*k); % p(k) obtained from p(k-1)
    e = e + (.75*k - .5*m)*pk;   % profit selling kth paper
  end;                           % finish profit for k <= m
  for k=m+1:n                    % sum profit for k > m
    pk = (n-k+1)*pk*p/((1-p)*k); % p(k) is obtained from p(k-1)
    e = e + .25*m*pk;            % profit selling kth paper
  end;                           % finish profits for k >= m
E(m) = e;                        % profit buying m papers
end;                             % end experiment with m
plot(E)                          % plot profit vs.papers bought
```

The routine above returns a plot of expected profit E versus the number m of papers bought by the newsboy. Profit maximizes at $m = 96$ papers, with a profit of \$22.32.

§1.5 The Poisson Distribution

On average, λ random noise spikes occur on channel per unit time. What is the probability that exactly k spikes will occur in a time period of length T?

If we partition the time interval $[0, T]$ into n equal subintervals, then the probability that a spike occured in k subintervals but not in any of the remaining $n - k$ other subintervals is

$$\binom{n}{k}\left(\frac{\lambda T}{n}\right)^k \left(1 - \frac{\lambda T}{n}\right)^{n-k} = \frac{n(n-1)\cdots(n-k+1)}{(n-\lambda T)^k}\frac{(\lambda T)^k}{k!}\left(1-\frac{\lambda T}{n}\right)^n.$$

So as the number of partitions n grows, the actual probability that k spikes occur is seen to be

$$p(k) = \frac{(\lambda T)^k}{k!}e^{-\lambda T}. \tag{1.25}$$

This is the discrete *Poisson* distribution with cumulative distribution function

$$F(x) = e^{-\lambda T} \sum_{0 \le k \le x} \frac{(\lambda T)^k}{k!} \tag{1.26}$$

with mean

$$\mu = \lambda T \tag{1.27}$$

and standard deviation

$$\sigma = \sqrt{\lambda T} \tag{1.28}$$

(Exercise 1.9). The Poisson distribution well models "arrival" problems.

Problem. A market has 60 customers per hour on average. What is the probability that more than 20 people will wish to check out within a given quarter-hour period?

Solution. (We ignore the dinner rush-hour peak and early morning lull.) Customers arrive at checkout with density $\lambda = 60$ people per hour. Thus the probability p that $k > 20$ will arrive at checkout in a particular period of length $T = 0.25$ hour is

$$p = e^{-15} \sum_{k=21}^{\infty} \frac{15^k}{k!} = 1 - q,$$

where

$$q = e^{-15} \sum_{k=0}^{20} \frac{15^k}{k!}. \tag{1.29}$$

The finite sum for q can be computed easily via MATLAB:

```
% Routine 1.3
q = 0;              % initialize q
m=20;               % number to arrive less 1
lT = 15;            % arrivals per quarter hour
p = 1;              % zeroth term of the sum
for k=1:m           % sum loop begins
  p = p*lT/k;       % kth term from the (k-1)th
  q = q + p;        % sum kth contribution
  end;              % end summation loop
q = q*exp(-lT);     % normalize
1-q                 % print p
```

This routine returns the probability of $p = 0.083$, i.e., less than a 9% chance of more than 20 wanting to check out within the quarter hour.

In Chapter 2 we examine this problem further by computing average wait time at checkout and optimizing the number of checkout counters. These are central questions for traffic on communication networks.

The Newsboy Problem Revisited

Poisson Solution. Since the numbers of passersby is large, this problem is well modeled by the Poisson distribution with $T = 1$ (one day) and hence by (1.27), $\lambda T = \mu = 100$. We must find when the expected profit

$$E_m = \sum_{k=0}^{m}(0.75k - 0.5m)\frac{e^{-100} \cdot 100^k}{k!} + \sum_{k=m+1}^{\infty} 0.25m \frac{e^{-100} \cdot 100^k}{k!}$$

maximizes. That is, when does

$$E_{m+1} - E_m = \text{(Exercise 1.14)} = 0.25 - 0.75 \sum_{k=0}^{m} \frac{e^{-100} \cdot 100^k}{k!} \quad (1.30)$$

first become negative? A slight modification of Routine 1.3 (Exercise 1.15) returns that the optimal number of papers to buy is $m = 96$.

§1.6 Taguchi Quality Control

Shutting down a production line because the product no longer meets specification is a drastic and expensive measure. How does a manufacturer design a production line to ensure that in spite of the normal variations in components, environment, and technique, the product produced will still be acceptable? The Japanese production engineer Genichi Taguchi proposed in 1986 the now widely accepted *quality loss function* (QLF) approach. Rather than being satisfied as long as a certain percentage of the product falls within given tolerance, manufacturers continually strive to reduce even a small QLF to lower values. This asymptotic reaching for perfection has had a profound effect on quality control.

One value of a measurement X of the product is set as the *target value* θ. Then taking into account "not only the immediate economic loss incurred, but also any long-term losses related to such factors as eventual loss of business to lack of goodwill or loss of competitiveness" [Derman and Ross], define a QLF

$$L(X, \theta) = k(X - \theta)^2. \qquad (1.31)$$

Note that the expected loss is

$$E[L(X, \theta)] = \cdots = k\sigma^2 + k(\mu - \theta)^2 \qquad (1.32)$$

(where as always μ and σ are the mean and standard deviation of the random variable X), thus agreeing with the intuition that quality is achieved by both meeting the targets on average and with small variation. As an epigram, *quality is increased by approaching the target and/or by decreasing the variance.*

Problem. One production approach yields items with a mean of $\mu = 1.5$ and a standard deviation $\sigma = 0.1$, while a second yields $\mu = 1.2$ with $\sigma = 0.4$. If the target is $\theta = 1.2$, which approach is best?

Solution. The expected loss of quality is proportional to $\sigma^2 + (\mu - \theta)^2$. So the first approach yields a loss of $(0.1)^2 + (1.5 - 1.2)^2 = 0.10$ while the second approach yields the higher loss of $(0.4)^2 + (1.2 - 1.2)^2 = 0.16$. The first approach is superior to the second even though the second exactly meets the target on average.

Problem. A company manufactures crystal oscillators with an advertised frequency temperature instability of at most ± 2 parts per million (ppm) in the temperature range $-20°$ to $+70°$C for a cost of \$340 each. Testing of each unit involves instrumenting and observing the unit for one workweek at an estimated additional cost of \$18 per unit. Previous testing has revealed that these units meet the specification on average with a standard deviation $\sigma = 0.5$ ppm. Should all units be tested with the costs passed on to the customer, thereby losing some competitive advantage from the higher unit cost?

Solution. At first glance the situation is rosy — all of the units out to 4 σ's meet the specification. But the Taguchi approach reveals the flaw.

Take as target $\theta = 0$ ppm temperature drift, giving the quality loss function $L(X) = kX^2$. At first cut, a failure to meet the ± 2 ppm specification results in a loss of \$340. Thus $k = 340/2^2 = \$85$ per $(\text{ppm})^2$. But then by (1.32) the expected loss is

$$E[L(X)] = k\sigma^2 + k(\mu - 0)^2 = k(0.5)^2 + 0 = \$21.25,$$

and thus weeklong testing is warranted. What is certainly warranted is an effort to reduce the variance $\nu = \sigma^2$ as well as a more careful determination of the loss factor k.

Cynics have derided the Taguchi QLF as a (admittedly successful) ploy to trick management into making sound engineering decisions by using the only units they understand — dollars.

Question. Before even setting up the production line, how do we identify the factors that most influence loss? Is it temperature during production? The purity of the solution? Which out-of-spec components most strongly increase loss of quality?

Knowing these sensitivities would allow us to concentrate our quality improvement efforts in the most cost-effective directions before production begins. This is called *off-line quality control design.* Taguchi has married the QLF to factor analysis to reveal the factors with the most effect upon the quality of the final product. The method is a variation on factorial design using *orthogonal arrays.*

Suppose that the quality of the product is determined by n production factors $x_1, x_2, \ldots, x_n$ within our control. Each factor x_j can be set anywhere within the range $a_j \leq x_j \leq b_j$. Our task is to discover which factor has the most effect on the expected loss of quality $E[L(X)]$. Let us normalize each factor with the change of variables

$$y_j = \frac{2x_j - (a_j + b_j)}{b_j - a_j} \tag{1.33}$$

so that the new normalized factors y_j range over $-1 \leq y_j \leq 1$.

Expand out the expected loss in a Taylor series

$$E[L(X)] = L_0 + \sum_{j=1}^{n} \alpha_j y_j + \sum_{i,j=1}^{n} \alpha_{ij} y_i y_j + \sum_{i,j,k=1}^{n} \alpha_{ijk} y_i y_j y_k + \cdots. \tag{1.34}$$

Let us first assume that the linear terms suffice, i.e.,

$$E[L(X)] \approx L_0 + \sum_{j=1}^{n} \alpha_j y_j \tag{1.35}$$

to good approximation. Our task has been reduced to finding the coefficient α_j with maximum absolute value, for quality will be most sensitive to the corresponding factor x_j.

Here is how an experiment is designed to discover this most important factor x_j. Find an $m \times n$ *orthogonal array* A such that:

1) Each entry is either $+1$ or -1.

2) Each column adds to 0.

3) The dot product of each column with every other column is 0.

Each row of A now dictates an experiment! For each $i = 1, 2, \ldots, m$, repeatedly perform a preproduction run with the factors set high, $x_j = b_j$, or low, $x_j = a_j$, depending on the setting in the jth column of the ith row of A, high if $+1$, low if -1. Record the average loss outcome L_i. Then, from (1.35),

$$A \begin{vmatrix} \alpha_1 \\ \alpha_2 \\ \cdot \\ \cdot \\ \cdot \\ \alpha_n \end{vmatrix} \approx \begin{vmatrix} L_1 - L_0 \\ L_2 - L_0 \\ \cdot \\ \cdot \\ \cdot \\ L_{n+1} - L_0 \end{vmatrix}. \tag{1.36}$$

But then because $A^T A = m I_{n \times n}$, and because columns of A sum to 0,

$$\begin{vmatrix} \alpha_1 \\ \alpha_2 \\ \cdot \\ \cdot \\ \cdot \\ \alpha_n \end{vmatrix} \approx \frac{1}{m} A^T \begin{vmatrix} L_1 - L_0 \\ L_2 - L_0 \\ \cdot \\ \cdot \\ \cdot \\ L_{n+1} - L_0 \end{vmatrix} = \frac{1}{m} A^T \begin{vmatrix} L_1 \\ L_2 \\ \cdot \\ \cdot \\ \cdot \\ L_m \end{vmatrix}. \tag{1.37}$$

The largest $|\alpha_j|$ is obtained from the signed averages on the right of (1.37). The corresponding factor x_j is the most critical.

Example. We have three factors and have repeatedly performed the four experiments with the displayed average outcomes as shown in Table 1.1. Thus from (1.37)

$$\alpha_1 = \frac{3.567 + 2.145 - 1.678 - 2.564}{4} \approx 0.3675$$

$$\alpha_2 = \frac{3.567 - 2.145 + 1.678 - 2.564}{4} \approx 0.1340$$

$$\alpha_3 = \frac{3.567 - 2.145 - 1.678 + 2.564}{4} \approx 0.5770,$$

so the manufacturing process is most sensitive to the third factor.

Table 1.1

Experiment	Factor 1	Factor 2	Factor 3	Average loss
1	1	1	1	3.567
2	1	−1	−1	2.145
3	−1	1	−1	1.678
4	−1	−1	1	2.564

In short, we need to perform repeatedly only m experiments instead of the naive 2^n; e.g., for $n = 7$ factors, only $m = 8$ experiments are necessary rather than 128. As for second- and higher- order effects in (1.34), we may again use this approach to test the size of the joint coefficients α_{ij}, α_{ijk}, and so on — see [Ryan].

Question. For each n, what is the smallest (even) $m > n$ for which an $m \times n$ orthogonal array exists?

This question has a hoary history, long associated with the name Hadamard. In fact, $n + 1 \equiv 0 \bmod 4 \Rightarrow m = n + 1$ is a famous conjecture [Hall]. You are guaranteed a $n + 1$ by n orthogonal array when $n + 1$ is a power of 2. See Exercises 1.18 through 1.22.

For an easy read on practical statistics see [Spiegel]. For the integration theory underlying probability, see [Cramér] or [Fabian and Hannan]. Morse and Kimball's classic first book on operations research is not for the squeamish since it is solely about warfare. More business-oriented books are [Wagner] and [Hillier]. We have only touched on Taguchi's ideas; for further quality control reading see [Taguchi and Wu], [Ryan], and [Derman and Ross].

Exercises

1.1 Repeatedly roll two fair dice. Graph the cumulative distribution function $F(x)$ for the random variable X that sums their top face values on each roll.

1.2 If x and y are drawn randomly from the interval $[0, 2]$, what is the probability that $x^2 + y < 1$?

Answer: 1/6.

Hint: How likely is a random point in the square $0 \leq x, y \leq 2$ to fall in the area under the curve $y = 1 - x^2$?

1.3 Give an example of a continuous cumulative distribution function $F(x)$ whose mean μ is quite different from its *median m* defined by $F(m) = 0.5$.

1.4 (**Project**) Sketch your best guess of the cumulative probability density function $f(x)$ of the age x of death of an American male. Carefully redraw the graph after consulting the *Statistical Abstracts of the United States*(an annual publication of the U.S. Bureau of the Census) or the Web. Propose a polynomial or spline probability density function to fit $f(x)$.

1.5 Why are the first several pages of a table of logarithms more worn than the last several?

1.6 Establish (1.12) and (1.13).

1.7 Show that except for one rare case, the sum of two uniformly distributed random variables is not uniformly distributed.

1.8 Establish the mean (1.20) and standard deviation (1.21) for the binomial distribution.

1.9 Establish the formulas for the mean (1.27) and standard deviation (1.28) of the Poisson distribution.

1.10 Show that if X and Y are independent and normally distributed, then $Z = \max(X, Y)$ is necessarily not normally distributed.

1.11 On average, one sodium vapor tunnel light burns out every week. How often each year will two or more lights burn out in 1 day?

1.12 The binomial solution of the newsboy problem in §1.3 does not seem to take into account the degree of clustering of the number of customers about the mean of 100. If the standard deviation is very small so that exactly 100 potential customers are essentially guaranteed, then 100, not 96 papers should be purchased. How do you escape this conundrum?

1.13 A normally distributed random variable X has mean $\mu = 72$ and standard deviation $\sigma = 11$. Find the probability that X lies between $a = 54$ and $b = 75$.

1.14 Establish (1.30).

1.15 Write a routine to verify via (1.30) that the Poisson solution to the newsboy problem of §1.4 is indeed $m = 96$.

1.16 Give an example of two normally distributed production outcomes X_1 and X_2, both of which to three σ's fall within a given tolerance band about a target $\theta = 0$, with drastically different expected quality losses (1.32).

1.17 Suppose that an item costs \$117 to manufacture and must be discarded if it misses the target of 12 by a margin of 3. Thus the QLF $L(X) = k(X - 12)^2$, where $k = 117/3^2 = 13$. The production methods have in the past yielded a mean of 11.7 and a standard deviation of 1.0. It has been discovered that an additional effort of \$20 per item can lower the standard deviation to 0.95. Is this a worthwhile effort?

1.18 As an intuition building exercise, attempt (futilely) to construct an orthogonal array that is 6×4.

1.19 Show that if A is an $m \times n$ orthogonal array, then $m \equiv 0 \bmod 4$.

1.20*Prove that when $n+1$ is a power of 2, there exists an $(n+1) \times n$ orthogonal array.

Outline: Recursively form $F_{k+1} = F \otimes F_k$, where

$$F = \begin{pmatrix} 1 & 1 \\ 1 & -1 \end{pmatrix}$$

and where '$\otimes$' is the Kronecker (tensor) product. Strike out the first column of F_{k+1} to obtain the required orthogonal array.

(We will again visit with these matrices F_k when we take up the fast Fourier transform.)

1.21 Construct an 8×7 orthogonal array.

1.22 Prove that in any $m \times n$ orthogonal array, $m > n$.

Hint: The columns are independent.

1.23 Requiring each entry of an orthogonal array to be either $+1$ or -1 is not crucial. Construct a 9×4 orthogonal array using an equal number of 1's, 0's, and -1's per column: where columns sum to 0 and where columns are orthogonal — thereby allowing the testing of four three-level factors.

1.24 MATLAB has a built-in Gaussian random number generator **randn** with mean $\mu = 0$ and standard deviation $\sigma = 1$. Test this random variable by constructing an incidence histogram; i.e., divide $[-4, 4]$ into say 80 subintervals (bins), call **randn** 1000 times, and count the number of times the variable falls into each bin. Normalize by the number of calls and graph against (1.15).

1.25 (**Project**) Go to a discount store and record the time of arrival of customers at the checkout counters. Prepare a graphical presentation of the cumulative probability distribution $F(t)$ and the probability density function $f(t)$ of the random variable T measuring time between arrivals. If help is available, also take statistics on the wait time in line and on time spent with the cashier for use in Chapter 2.

1.26 Suppose that X is Gaussian noise with mean $\mu = 0$ and standard deviation $\sigma = 1$. Find the cumulative distribution function $G(y)$ of the noise Y obtained by passing X through an *ideal rectifier*, i.e., $Y = X$ if $X \geq 0$ but $Y = 0$ if $X < 0$.

1.27 Suppose that X is Gaussian noise with mean $\mu = 0$ and standard deviation σ. Find the cumulative distribution function $G(y)$ of the noise Y obtained by passing X through a *square-law detector*, i.e., $Y = X^2$.

1.28 Suppose that we are firing at a target marked in the usual way with concentric circles of increasing radius r. Our error is Gaussian and independent in the horizontal and vertical directions with mean $\mu = 0$ and standard deviation σ. Show that the error in terms of the distance r from the center of the bull's-eye has the *Rayleigh* cumulative distribution function

$$F(r) = 1 - e^{-r^2/2\sigma^2}.$$

1.29 A train leaves the depot empty. At each station, on average, 40 people mount while 20% of the riders dismount. What is the expected number of people on the train after the nth stop?

1.30 (**Project**) Design an overbooking strategy for an airline. Suppose that each passenger has probability $p < 1$ of showing up for one of the N seats on a given flight. Each empty seat represents a loss of L dollars. What maximum reparation P should you offer each of the m surplus ticketed passengers to stand down from an overbooked flight? Using the binomial distribution, balance the expected loss from flying empty seats with the expected cost of reparations. Play off m against P. What strategy most reduces loss of perceived quality?

1.31 Draw a random chord L of a circle C. What is the probability that the length of L exceeds the side length of an inscribed equilateral triangle? Find at least two plausible but different probability models that yield distinct answers. Write a persuasive argument for each model.

1.32 (**Project**) The Michigan Lotto Commission publishes the following "Odds of winning: Match 6 of 6: 1 in 13,983,816; Match 5 of 6: 1 in 54,201; Match 4 of 6: 1 in 1,032. Overall odds: 1 in 1,013." How are these numbers obtained? Assuming that 1 million play, how large must the prize be before a \$1 bet can expect to return more than \$1?

1.33 Why do physical constants lead with the digit 1 more often than any other digit?

1.34*Argue intuitively, then measure theoretically, that the cumulative distribution function $y = F(x)$ of (1.1) is continuous from the right.

1.35*We saw in §1.5 that when $p/n = \lambda T$ is held constant as n grows large, the binomial distribution becomes Poisson in the limit.

In contrast, prove that if p is held fixed as $n \to \infty$, the binomial distribution becomes Gaussian.

1.36*Let X be a Gaussian random variable with mean μ and standard deviation σ. By completing the square, show that the expected value of the random variable $Y = e^{aX}$ is

$$E[e^{aX}] = e^{a\mu + a^2\sigma^2/2}.$$

1.37 Are birthdays uniformly distributed throughout the year?

1.38 Supposing that birthdays are uniformly distributed throughout the year, what is the probability that at least two people in a class of 30 were born on the same day? How large must this class be to guarantee a repeated birthday with a confidence of 90%?

1.39 A function f is drawn from the bag of all functions that map a set X of m elements into a set Y of n elements. What is the probability that f is injective?

1.40*If m people enter an elevator, each pressing a button for a floor, what is the probability that all n buttons will be pressed?

1.41*Assume that each particle of a system can exist at only one of the discrete energy levels $E_1 < E_2 < E_3 < \cdots$. Let p_n be the frequency with which a particle is found at level E_n, giving that $1 = p_1 + p_2 + p_3 + \cdots$. Fix the mean energy of the system $E = E_1 p_1 + E_2 p_2 + E_3 p_3 + \cdots$. Show that when *entropy* $H = -p_1 \log p_1 - p_2 \log p_2 - p_3 \log p_3 + \cdots$ is maximum, there is a unique *temperature* T so that each $p_n = e^{-E_n/T}/Q$ where the *partition function* $Q = e^{-E_1/T} + e^{-E_2/T} + e^{-E_3/T} + \cdots$ and hence $H = E/T + \log Q$ ([Feynman]).

Hint: Use Lagrange multipliers.

Chapter 2
The Monte Carlo Method

*The computer repeatedly experiments to build a statistical
picture of the phenomenon.*

Often, it is not cost-effective or even possible to obtain an analytic
model of a phenomenon. Instead, we simulate the mechanism, in-
structing the computer to randomly try various inputs and to tabulate the results. This statistical data is often enough to answer the
question posed. We begin with "dartboard" integration followed by
a reliability study of multicomponent manufactured products. We
then take up the servicing of requests, a problem of dire importance
in large communication networks. We finish with a final look at our
newsboy at work.

§2.1 Computing Integrals

Our goal is to compute the area below the curve $y = f(x)$ shown in
Figure 2.1.

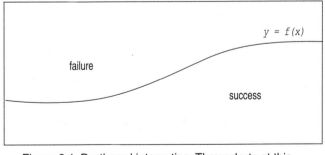

Figure 2.1. Dartboard integration. Throw darts at this
rectangular target. The integral is the success rate
times the target area.

Repeatedly choosing a point at random within the rectangle enclosing the curve, we keep count of the number of times this random point falls under the curve (success). After a large number of experiments are performed, we multiply the area of the enclosing rectangle by the success rate to obtain an estimate of the area under the curve.

Example. Let us estimate the value of the integral

$$I = \int_0^1 e^{-x^3} \, dx. \tag{2.1}$$

Draw the box $0 \le x \le 1$ and $0 \le y \le 1$ to enclose the curve. We now exploit the built-in random number generator **rand** of MATLAB that will generate uniformly distributed random numbers between 0 and 1:

```
% Routine 2.1
A = 1;                      % area of enclosing rectangle
N = 10000;                  % set the number of trials
s = 0;                      % initialize success counter
for i = 1:N                 % begin trials
  x = rand;                 % choose a random x-coordinate
  y = rand;                 % choose a random y-coordinate
  if y <= exp(-x^3)         % if below the curve, then
    s = s + 1;              % increment successes.
   end;                     % end if
  end;                      % end of each trial
I = A*s/N                   % integral=area*success/trials
```

This routine yields an estimate for (2.1) of $I \approx 0.805$.

We have bowed to tradition by introducing the Monte Carlo method as a method of integration. In truth, for functions of one variable, there are superior methods of numerical integration, such as Simpson's rule. Nevertheless, important and practical uses of Monte Carlo often involve the evaluation of difficult integrals that are hidden from view because there is no explicit analytic model. There is no better example of this than the next application.

§2.2 Mean Time Between Failures

Suppose that we make a product composed of n components where the ith component is guaranteed by its manufacturer to possess a time before failure that is normally distributed about the mean μ_i

with standard deviation σ_i. Assume that our product will fail when any one of its components fail.

Question. What guarantee can we offer our customers for our product? What is our product's mean time between failures (MTBF)?

Analytic Solution. We must compute the mean of the random variable

$$T = \min(T_1, T_2, \ldots, T_n), \tag{2.2}$$

where T_i is the time to failure of the ith component. If $F_i(t)$ is the cumulative distribution function (CDF) of the ith component, then the CDF $F(t)$ of the random variable T is

$$F(t) = \mathrm{prob}(T \le t) = 1 - \mathrm{prob}(T > t) \tag{2.3}$$

$$= 1 - \mathrm{prob}(\min(T_1, T_2, \ldots, T_n) > t)$$

$$= 1 - \mathrm{prob}(T_1 > t \text{ and } T_2 > t \text{ and } \cdots \text{ and } T_n > t) \tag{2.4}$$

$$= 1 - \prod_{i=1}^{n} [1 - F_i(t)] \tag{2.5}$$

giving that the MTBF of our product is

$$\mu = \sum_{i=1}^{n} \int_{-\infty}^{\infty} t \prod_{j \ne i} [1 - F_j(t)] \, dF_i(t), \tag{2.6}$$

a hopeless computational task.

Monte Carlo Solution. Let us merely watch this product fail over and over via computer simulation, then average the failure times. To keep ideas definite, let us assume that we have three components with MTBF 11, 12, and 13 years with standard deviations of 1, 2, and 3 years respectively.

We exploit the built-in Gaussian number generator **randn** of MATLAB, which generates normally distributed values with mean $\mu = 0$ and standard deviation $\sigma = 1$:

```
% Routine 2.2
N = 10000;                  % set the number of trials
m = 0;                      % initialize mean failure time
m2 = 0;                     % initialize the second moment
for i = 1:N                 % begin trials
```

```
x = 11 + randn;              % fail time for component 1
y = 12 + 2*randn;            % fail time for component 2
z = 13 + 3*randn;            % fail time for component 3
w = min([x,y,z]);            % min.  fail time of all three
m = m + w/N;                 % add trial to mean fail time
m2 = m2 + w*w/N;             % add trial to second moment
 end;                        % end of each trial
m                            % display the MTBF
sd = sqrt(m2 - m*m)          % compute the standard deviation
```

This routine gives the MTBF as $\mu \approx 10.12$ years with standard deviation $\sigma \approx 1.38$.

Warning. In practice, multicomponent systems experience a high "infant mortality rate." Due to manufacturing/maintenance errors and/or shipping/installation mishaps, the frequency of failure during the burn-in period is higher than the mean time *between* failures would suggest. See [Frankel, p. 17] and [Clements]. The likelihood of failure with time is a U-shaped curve.

§2.3 Servicing Requests

Suppose that we wish to predict typical checkout time at a discount store, or perhaps anticipate a server's response time to arriving packets. Assume that the time T between requests for service (arrival time) obeys a *Poisson flow* probability density function

$$f(t) = \frac{e^{-t/a}}{a}, \qquad t \geq 0, \ a > 0, \qquad (2.7)$$

where $f(t) = 0$ for negative t (Exercise 2.14). Assume that we have m service lines (or processors) to handle these requests. If the first processor is busy, the request is handed off to the second, and so on. If all processors are busy, the request is rejected. Let us assume that all of the m processors service each request in time S, where S is normally distributed with mean μ and standard deviation σ.

Question. On average, how many requests are handled in a given time period, and how many are rejected?

Solution. Let us experiment with $m = 3$ processors, each with servicing time S with $\mu = 20$ and $\sigma = 1$. Requests arrive in intervals of time T in (2.7) determined by $a = 10$ so that the mean arrival time is 10 (Exercise 2.10). But how do we experimentally generate these arrival times T?

LEMMA. Suppose that the random variable X has the cumulative distribution function $F(x)$. Suppose also that $F(x)$ is everywhere continuous and strictly increasing whenever it is nonzero and less than 1. Then $X = F^{-1}(Y)$, where Y is uniformly distributed on $[0, 1]$.

PROOF. Exercise 2.11.

By this lemma we can construct the random variable T of arrival times with distribution (2.7) by solving for t in

$$y = F(t) = \int_0^t f(s)\, ds = \frac{1}{a} \int_0^t e^{-s/a}\, ds \qquad (2.8)$$

in terms of y. Thus (Exercise 2.12)

$$T = -a \ln(1 - Y) \qquad (2.9)$$

is distributed as in (2.7) when Y is uniformly distributed $[0, 1]$, a distribution available in MATLAB as **rand**. We are now ready to perform Monte Carlo experiments.

```
% Routine 2.3
a =10;                    % arrival time parameter
m = 3;                    % number of processors
mu = 20;                  % mean processing time
sig = 1;                  % s.d. of the processing time
Q = 1:m;                  % set the # processors = m
Q = 0*Q;                  % initialize the time when free
t = 0;                    % set the time clock to zero
rej = 0;                  % set reject count to 0
N = 10000;               % watch for N arrivals

for k=1:N                 % begin the experiment
 busy = 1;               % set ``all busy'' toggle
 T = - a*log(1-rand);    % arrival time of new customer
 t = t + T;              % update time clock to present
 for i=1:m               % check to see who's busy:
 if Q(i) < t             % if i is free, then
  S = mu+sig*randn;      % compute time to service,
  Q(i) = t+ S;           % put i to servicing the request,
  busy=0;                % and signal no reject necessary,
```

```
    break;                      % escape to wait for next arrival.
    end;                        % endif
  end;                          % stop canvasing processors
if busy==1                      % if all processors are busy, then
  rej = rej+1;                  % increment the number of rejects.
  end;                          % endif
end;                            % end experiment
rej/N                           % print the ratio of rejects
```

The routine shows that approximately 21% of the requests are rejected. But a grocery store cannot reject requests from customers wishing to check out.

Question. If requests cannot be rejected, what is the average wait time in line?

Solution. We assume that the customer will search out the checkout lane with the shortest wait (least number of items in the carts), not the line with the least number of people. Suppose that we have $m = 10$ checkout lanes, arrival time T given by (2.7) with mean 1 minute, i.e., $a = 1$, and that checkout time S is normally distributed with mean $\mu = 8$ and standard deviation $\sigma = 3$ minutes.

```
% Routine 2.4
a = 1;                          % arrival time parameter T
m = 10;                         % number of cashiers
mu = 8;                         % mean processing time
sig = 3;                        % s.d.  of processing time
Q = zeros(1,m);                 % establish cashier's backlog
wait=0;                         % initialize total wait time
N = 10000;                      % experiment with N customers
                                % store opens
for k=1:N                       % watch N customers
  T = -a*log(1-rand);           % arrival time of new customer C
  S = mu + sig*randn;           % compute time to service C
  Q = max(Q - T, 0);            % update the queues to present
  [t,j] = min(Q);               % j has shortest wait time
  wait = wait + Q(j);           % add C's time to reach cashier
  Q(j) = Q(j) + S;              % time for C to clear queue j
end;                            % store closes
wait/N                          % average time to reach cashier
```

The experiment yields an average time to reach a cashier of a little less than 1 minute.

§2.4 The Newsboy Problem (Reprise)

Recall from §1.4 that our intrepid newsboy with 100 customers on average must pay \$0.50 for each paper (with no returns) and that he sells at \$0.75. A binomial model and a Poisson model both support that maximal profit is obtained when only 96 papers are purchased. Rather than the rather complicated calculations of (1.4) and Routine 1.2, a Monte Carlo approach is direct and simple — we watch the newsboy at work for, say, one year, experimenting with the number of prepurchased papers:

```
% Routine 2.5
N = 5000;                              % number of passersby per day
p = 100/N;                             % prob.  each is a customer
d = 365;                               % experiment for a year
papers = 90:105;                       % try buying 90 to 105 papers
prof = 90:105;                         % size the profit vector and
prof = 0*prof;                         % initialize to 0.
sold = prof;                           % size the # sold vector
                                       % begin the experiment
for i=1:d                              % loop through the year
  sold = 0*sold;                       % none sold at start of day
  for k=1:N                            % people begin walking by
    if rand <= p                       % if he/she is a customer,
    sold = sold+(sold<=papers);        % sell a paper if any are left
    end;                               % endif
  end;                                 % day ends
  prof=prof+.75*sold-.5*papers;        % add in day's profit
end;                                   % end year
prof = prof/d;                         % average daily profit
plot(papers,prof)                      % plot average daily profit
```

The experiment shows maximal profit when buying $m = 95$ or 96 papers.

For further reading on the Monte Carlo method see the short primer by I.M. Sobol. The classic on serving requests is [Hillier].

Exercises

2.1 Using Monte Carlo, estimate the value of

$$\int_0^1 x^2 \, dx.$$

Experiment with the number of trials. Graph accuracy against number of trials.

2.2 Modify Routine 2.1 to estimate the integral of $f(x) = 5\sin^2 x$ for $0 \le x \le \pi$.

2.3 Using Monte Carlo, estimate the value of

$$\int_0^1 \int_0^1 e^{xy} \, dx \, dy.$$

2.4 How would you modify the Monte Carlo method in order to integrate not necessarily positive functions? Adapt Routine 2.1 to estimate

$$\int_0^{2\pi} \frac{\sin x}{x} \, dx.$$

2.5 Why does (2.5) follow from (2.4)?

2.6 Establish (2.6). Write out explicitly what integration must be done for the case $n = 2$.

2.7 Suppose that a product composed of two parts — each with MTBF of 10 years and standard deviation 1 — will fail if either component fails. Explain in layman's terms why this construct has a shorter life than its components. Specialize the code of Routine 2.2 to show that on average the product will fail in 9.4 years.

2.8 A manufactured product consists of three components, all with MTBF $\mu = 10$, $\sigma = 1$. The first component is a backup for the second; i.e., either both must fail or the third component must fail before the construct fails. Find the MTBF μ.

Answer: $\mu \approx 9.7$.

2.9 Write a general MTBF routine for n components in series, i.e., where the construct fails whenever any component fails. Run the case $n = 12$, where the ith component has $\mu_i = i$ and $\sigma_i = 1/i$.

2.10 Show that the mean and standard deviation of the Poisson arrival (2.7) is $\mu = \sigma = a$.

2.11 Explain why the lemma of §2.3 is geometrically obvious. Then carefully construct an analytic proof.

2.12 Establish (2.9).

2.13 Experiment with Routine 2.3. Graph the rate of rejections against mean arrival time a, then against the standard deviation σ of service time.

2.14 (**Project**) Why is the Poisson flow in (2.7) a reasonable guess for arrival times? Write a paragraph arguing that this flow is a correct model. Then write a paragraph arguing that it is not the correct model. Next visit a discount store and time the arrival of customers at checkout. Also time the arrival of customers at one particular checkout lane. Does either set of data appear to match the Poisson flow?

2.15 (**Project**) Experiment with Routine 2.4. Graph mean wait against mean service time μ. Write additional code to find how long cashiers are idle on average. Think on how this business could be optimized — play off customer satisfaction against costly cashier idleness. Use Taguchi's QLF (§1.6).

2.16*Revise Routine 2.4 so that each customer chooses the checkout lane with the fewest number of people in line rather than the least apparent wait. Are your results substantially different than those of Routine 2.4?

2.17*Every day you must give your dog one half of a heartworm pill. You begin the mosquito season with a large bottle of N whole pills. Each day you decant an object from the bottle: If it is a half pill, give it to the dog; if whole, break it in half, administer one half to the dog, and return the other half to the bottle. After k days, what is the expected number of unbroken pills remaining?

2.18 (Project) Let X be the random variable

$$X = \frac{X_1 + X_2 + \cdots + X_N}{\sqrt{N}}$$

where X_k is the outcome from the kth flip of a fair coin where heads yields the value 1 and tails -1. Show $E[X] = \mu = 0$ and $\sigma^2 = E[X^2] = 1$. But what is its distribution? Show experimentally that for large N, X appears to be normally distributed. You will be substantiating the celebrated *central limit theorem* stated in any book on statistics.

2.19 (Project) In the same vein as in Exercise 2.18, an approximately normal random variable with $\mu = 0$, $\sigma = 1$ is often constructed by adding together 12 independent, uniformly distributed random variables with support $[0, 1]$ and subtracting 6. Construct such a random variable using MATLAB's uniformly distributed **rand** and check its density function for normality.

2.20 Simulate a drunkard's walk. As he leaves the bar at $(0, 0)$, each step is in a random direction of length $L = 1$. Estimate the expected value of the square $r(N)^2$ of his distance $r(N)$ from the bar after N steps. Estimate the expected value of $r(N)$. Why is $E[r(N)^2] \neq E[r(N)]^2$?

Exact answer: $E[r(N)^2] = N$.

Outline: After the kth step,

$$(x_k, y_k) = (x_{k-1}, y_{k-1}) + (\cos\theta, \sin\theta)$$

where θ is chosen randomly from the interval $[0, 2\pi]$.

2.21 What if the drunkard's step length L is random with mean μ and standard deviation σ? Show that in this case,

$$E[r(N)^2] = (\sigma^2 + \mu^2)N.$$

What if the stagger angle θ is constrained by $|\theta| < \theta_0 < \pi$?

Chapter 3

Data Acquisition and Manipulation

A new and rich body of mathematics has arisen to deal with discrete data streams.

We see here how to exploit the discrete data taken by computers by transforming data streams into the *frequency domain* with the z-transform, the discrete analog of the Laplace transform. This transform provides an effective method for solving linear recursions. You will see how to design filters to yield more useful data and how to judge their performance with Bode plots. You will learn how to spot instability in mechanisms by checking a simple criterion. A fundamental flaw of sampling — *aliasing* — will be revealed. The chapter ends with an introduction to feedback control theory via the design of an automobile cruise control plus practice with decibel calculations.

§3.1 The z-Transform

Suppose that we are monitoring the progress of some scientific or industrial process, say the temperature of a reaction, the movement of a piston, or the voltage on a port. While the process is almost certainly continuous, the computer measuring the process is taking data only at (usually) equally spaced intervals T whose length is determined by the speed of the process — there may be no need to measure outside temperature every millisecond, but often a need to measure piston position at such high rates.

The typical apparatus is a sensor (transducer) that converts the process measurement to a voltage which is read by an *analog-to-digital* (A-D) converter, which is in turn debriefed periodically by a computer that smoothes, manipulates, stores, and displays the data (see Figure 3.1).

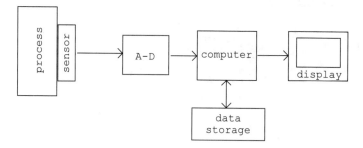

Figure 3.1. Data acquisition system.

Assume that the computer is taking data at intervals of T seconds, storing the sequence of measurements as (typically) 12- to 16-bit binary words:

$$x = \{x_0, x_1, x_2, x_3, \ldots\}. \tag{3.1}$$

But let us for now idealize the computer by granting it infinite precision, thinking of the data (3.1) as a sequence of real numbers as shown in Figure 3.2.

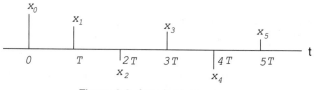

Figure 3.2. Acquired signal x .

This is the *time-domain* version of the acquired data x. We also think of this signal x as having an alter ego X in the *frequency domain* as the Laurent series

$$X = \sum_{k=0}^{\infty} x_k z^{-k} = x_0 + \frac{x_1}{z} + \frac{x_2}{z^2} + \frac{x_3}{z^3} + \cdots . \tag{3.2}$$

This frequency-domain object X is called the *z-transform* of the signal x, and we write

$$x \to X. \tag{3.3}$$

For example, the constant signal $\alpha, \alpha, \alpha, \alpha, \ldots$ has the z-transform

$$\sum_{k=0}^{\infty} \alpha z^{-k} = \frac{\alpha}{1 - z^{-1}} = \frac{\alpha z}{z - 1}, \tag{3.4}$$

while the signal of powers $1, \alpha, \alpha^2, \alpha^3, \ldots$ transforms to

$$\sum_{k=0}^{\infty} \alpha^k z^{-k} = \frac{1}{1 - \alpha z^{-1}} = \frac{z}{z - \alpha}. \tag{3.5}$$

In particular, a sine wave $x(t) = \sin \omega t$ sampled every T seconds has transform

$$\sum_{k=0}^{\infty} (\sin \omega Tk) z^{-k} = \sum_{k=0}^{\infty} (\alpha^k - \alpha^{-k}) z^{-k}/2i$$

$$= \text{(Exercise 3.1)} = \frac{\sin \omega T}{z - 2 \cos \omega T + z^{-1}}, \tag{3.6}$$

where $\alpha = \exp(\omega T i)$. Note that the z-transform is *linear*; i.e.,

$$\alpha x + \beta y \longrightarrow \alpha X + \beta Y \tag{3.7}$$

for any scalars α and β. Also note that delay of x by one sample, i.e., the signal $0, x_0, x_1, x_2, \ldots$ has the transform

$$\sum_{k=0}^{\infty} x_{k-1} z^{-k} = \frac{x_0}{z} + \frac{x_1}{z^2} + \frac{x_2}{z^3} + \cdots = z^{-1} X. \tag{3.8}$$

That is, *delay by one sample period is multiplication by z^{-1}*.

Example. Suppose we instruct the computer to take a weighted average of successive values of x to form a new, presumably smoothed signal

$$y_k = \alpha x_k + \beta x_{k-1}, \tag{3.9}$$

where $\alpha + \beta = 1$ and $x_{-1} = 0$. Then the z-transform of y is

$$Y = \sum_{k=0}^{\infty} y_k z^{-k} = \sum_{k=0}^{\infty} (\alpha x_k + \beta x_{k-1}) z^{-k}$$

$$= \alpha X + \beta z^{-1} X = (\alpha + \beta z^{-1}) X, \tag{3.10}$$

giving a hint of what is to come: *This averaging filter is in the frequency domain merely the product of the signal X with the fixed signal $\alpha + \beta z^{-1}$.*

This suggests a systematic method for solving linear recursion relations.

§3.2 Linear Recursions

We shall see that practical questions give rise to recursion relations of the form

$$x_k = \alpha_1 x_{k-1} + \alpha_2 x_{k-2} + \cdots + \alpha_n x_{k-n}, \tag{3.11}$$

where $\alpha_1, \alpha_2, \ldots, \alpha_n$ are fixed and given, and where the n initial values $x_{-1}, x_{-2}, \ldots, x_{-n}$ are known. The objective is to determine in closed form all future values x_k for $k \geq 0$. The z-transform provides an effective solution method: *Transform the recursion to the frequency domain, solve, and then transform the solution back to the time domain.*

Example. Let us solve the recursion

$$x_k = x_{k-1} + 2x_{k-2} \tag{3.12}$$

subject to the initial conditions $x_{-1} = 1/2$ and $x_{-2} = -1/4$. The rule is: Add the previous value to twice the value previous to that. So $x_0 = 1/2 + 2(-1/4) = 0$, $x_1 = 0 + 2 \cdot 1/2 = 1$, $x_2 = 1$, $x_3 = 3$, $x_4 = 5$, and so forth.

Let

$$X = \sum_{k=0}^{\infty} x_k z^{-k} \tag{3.13}$$

be our sought-for signal. Then the delayed signal

$$X_{-1} = \sum_{k=0}^{\infty} x_{k-1} z^{-k} = z^{-1} X + x_{-1} = z^{-1} X + 1/2 \tag{3.14}$$

and the doubly delayed signal

$$X_{-2} = \sum_{k=0}^{\infty} x_{k-2} z^{-k} = x_{-2} + \frac{x_{-1}}{z} + \frac{x_0}{z^2} + \cdots$$

$$= z^{-1} X_{-1} + x_{-2} = z^{-2} X + (1/2)z^{-1} - 1/4. \tag{3.15}$$

The recursion (3.12) becomes

$$X = X_{-1} + 2X_{-2}, \tag{3.16}$$

so our solution in the frequency domain is (Exercise 3.3)

$$X = \frac{z^{-1}}{1 - z^{-1} - 2z^{-2}} = \frac{z}{z^2 - z - 2}. \tag{3.17}$$

Expand in partial fractions

$$X = \frac{z}{(z+1)(z-2)} = \frac{A}{z+1} + \frac{B}{z-2} \qquad (3.18)$$

whereupon $A = 1/3$ and $B = 2/3$. Thus our solution

$$
\begin{aligned}
X &= \frac{1/3}{z+1} + \frac{2/3}{z-2} = \frac{1}{3z} \cdot \frac{1}{1+z^{-1}} + \frac{2}{3z} \cdot \frac{1}{1-2z^{-1}} \\
&= \frac{1}{3z} \sum_{k=0}^{\infty} (-1)^k z^{-k} + \frac{2}{3z} \sum_{k=0}^{\infty} 2^k z^{-k} \\
&= \sum_{k=0}^{\infty} [(1/3)(-1)^k + (2/3)2^k] z^{-k-1} \\
&= \sum_{k=1}^{\infty} [(1/3)(-1)^{k-1} + (1/3)2^k] z^{-k},
\end{aligned}
$$

giving the closed-form solution

$$x_k = (1/3)[2^k - (-1)^k]. \qquad (3.19)$$

Problem. We are terrorists who wish to gain influence on the world scene by constructing and detonating a small nuclear device. We begin with uranium hexafluoride of concentration $x_0 = 0.7\%$. Each of our centrifuge reactors in a chain (cascade) yields a product of improved concentration x_k over its input feedstock concentration x_{k-1} according to the classic chemical reactor equation

$$x_k = \frac{x_{k-1} + \tau\alpha}{1 + \tau(\alpha + \beta)}, \qquad (3.20)$$

where τ is holding time in the reactor, and where α and β are the reaction rates of the impurity and product, respectively. How many reactors N will we need to obtain the fissionable concentration $x_N = f = 5\%$ for a holding time of τ?

Solution. The reactor equation (3.20) is of the common form

$$x_k = ax_{k-1} + b \qquad (3.21)$$

with transform

$$X = az^{-1}X + ax_{-1} + \frac{b}{1 - z^{-1}} \qquad (3.22)$$

and solution

$$X = \frac{ax_{-1} + b - ax_{-1}z^{-1}}{(1 - az^{-1})(1 - z^{-1})}$$

$$= \frac{x_0 z^2 - ax_{-1}z}{(z-a)(z-1)} = x_0 + \frac{A}{z-a} + \frac{B}{z-1} \qquad (3.23)$$

$$= x_0 + \frac{A}{z}\sum_{k=0}^{\infty} a^k z^{-k} + \frac{B}{z}\sum_{k=0}^{\infty} z^{-k}$$

$$= x_0 + \sum_{k=1}^{\infty}(Aa^{k-1} + B)z^{-k}, \qquad (3.24)$$

giving that

$$x_k = Aa^{k-1} + B. \qquad (3.25)$$

We determine A and B with *Heaviside's left thumb:* Multiply both sides of (3.23) by $z - a$ and then set $z = a$ to obtain

$$A = \frac{x_0 a^2 - ax_{-1}a}{a-1} = \frac{x_0 a^2 - ax_0 + ab}{a-1} = ax_0 + \frac{ab}{a-1}. \qquad (3.26)$$

Multiply both sides of (3.23) by $z - 1$ and set $z = 1$ to obtain

$$B = -\frac{b}{a-1}. \qquad (3.27)$$

Writing out A and B in terms of α, β, τ from (3.20) and imposing the purity requirement $f = 5\%$ to fission, we may solve for the holding time τ (Exercise 3.9). Then we may play off the total time required to yield a critical mass m using reactors of capacity V against the marginal cost of additional reactors. For an optimization approach, see Exercise 3.10.

§3.3 Filters

What exactly is a filter? It is certainly some device F in hardware or software that yields an output more useful than its input. In this chapter we are interested in filters implemented within a computer by some algorithm: A data stream (signal) $u = \{u_k\}_{k=0}^{\infty}$ is passed through the filter F to yield the new signal $y = \{y_k\}_{k=0}^{\infty}$ according to some rule, satisfying for our present purposes the three requirements

1.) F is *linear*; i.e., if $y^{(1)} = F\,u^{(1)}$ and $y^{(2)} = F\,u^{(2)}$, then

$$F(\alpha u^{(1)} + \beta u^{(2)}) = \alpha y^{(1)} + \beta y^{(2)}, \qquad (3.28)$$

so that the principle of superposition obtains. (Sum and scalar product is termwise.)

2.) F is *causal*; i.e., the output cannot anticipate the input. In symbols, if $y = F\,u$, then

$$u_k = 0 \text{ for all } k < k_0 \text{ implies } y_k = 0 \text{ for all } k < k_0. \qquad (3.29)$$

3.) F is *time invariant*; i.e., its properties do not drift over time. In symbols, if $y = Fu$, then

$$F\left\{u_{k+k_0}\right\}_{k=0}^{\infty} = \left\{y_{k+k_0}\right\}_{k=0}^{\infty}. \qquad (3.30)$$

There are important nonlinear filters as well as filters that slowly adapt their parameters to changing conditions. Alas, there are no noncausal filters (see Exercise 3.11). But here and now we restrict our attention to linear causal time-invariant filters.

Question. Imagine that the computer is performing some filtering operation. What algorithm is it using?

Answer. To recover the algorithm, *ring the filter with the unit impulse;* i.e., input the signal $\delta = \{1, 0, 0, 0, 0, \dots\}$ and record the (causal) *impulse response* $h = \{h_0, h_1, h_2, \dots\}$.

RESULT A. The response $y = \{y_k\}_{k=0}^{\infty}$ of the linear causal time in-variant digital filter F to an arbitrary input $u = \{u_k\}_{k=0}^{\infty}$ is the discrete convolution of the input u with the impulse response h; i.e.,

$$y_k = u_k h_0 + u_{k-1} h_1 + \cdots + u_0 h_k = \sum_{j=0}^{k} h_j u_{k-j}. \qquad (3.31)$$

In the frequency domain $Y = HU$. Thus *filters act like power series multiply.*

PROOF. Write an arbitrary input signal u as a sum of delayed unit impulses:

$$u = \sum_{j=0}^{\infty} u_j \delta^{(j)},$$

where $\delta^{(j)} = \{0, 0, \ldots, 1(j\text{th position}), 0, 0, \ldots\}$. By linearity and time invariance,

$$y = Fu = \sum_{j=0}^{\infty} u_j F\delta^{(j)} = \sum_{j=0}^{\infty} u_j h^{(j)}$$

$$= u_0\{h_0, h_1, h_2, \ldots\} + u_1\{0, h_0, h_1, h_2, \ldots\} \qquad (3.32)$$
$$+ u_2\{0, 0, h_0, h_1, h_2, \ldots\} + \cdots,$$

so

$$y_k = u_0 h_k + u_1 h_{k-1} + \cdots + u_k h_0. \qquad (3.33)$$

Conversely, any causal signal h can be used as an impulse response to form the linear causal time-invariant filter $y = h * u$, where $*$ denotes the discrete convolution (3.31).

Example. The filter with the finite impulse response (FIR) $h = \{1/2, 1/2, 0, 0, 0, \ldots\}$ is the averaging filter

$$y_k = \frac{u_k + u_{k-1}}{2}, \qquad (3.34)$$

i.e., in the frequency domain

$$Y = \frac{1 + z^{-1}}{2} U. \qquad (3.35)$$

The z-transform $H(z)$ of the impulse response h is called the *transfer function* of the filter F.

Problem. How can we filter out the 60-Hz hum induced by the room wiring that is picked up by the electrodes of an electrocardiogram (EKG) machine?

Solution. Let us sample at say 720 Hz, i.e., $T = 1/720$ s, so that each 60-Hz sine wave is sampled 12 times. Then we may "notch out" the offending 60 Hz by adding the present input u to u delayed by six samples, thereby adding the 60-Hz component back in exactly 180° out of phase. In symbols,

$$y_k = u_k + u_{k-6}, \text{ i.e., } Y = (1 + z^{-6})U. \qquad (3.36)$$

Unfortunately, this (much too) simple filter will also notch out any sine waves of frequency an odd multiple of 60 Hz (Exercise 3.12), some of which may be highly diagnostic. We will do a better design below.

Problem. We wish to detect the presence or nonpresence of a faint sinusoidal signal deeply buried in galactic noise from a distant space-craft. Or more prosaically, we must detect low-level sensor responses that indicate the presence of metal shavings in a stream of oatmeal on its way to being boxed for sale.

Solution. First convert the signal to an IF (intermediate frequency) f_0 hertz low enough to permit digital signal processing (DSP) to take place. Sampling at N times per period, we wish to decide on the presence or nonpresence of $x_k = \sin 2\pi k/N$. It is a fundamental fact of DSP that the best filter for this purpose is the *matched filter,* where the impulse response h is an uncorrupted version of the sought-for signal itself [Papoulis]. Theory guarantees that using m periods of x as the terms of a FIR filter h, i.e.,

$$h_k = \begin{cases} \sin(2\pi k/N) & \text{if } 0 \le k \le mN \\ 0 & \text{otherwise,} \end{cases} \tag{3.37}$$

results in a predicted improvement in a signal-to-noise ratio (S/N) of approximately mB/f_0, where B is the IF bandwidth in herz [Papoulis]. In Exercise 3.14 you are asked to simulate this filter.

§3.4 Stability

We say that a filter F with impulse response $h = \{h_k\}_{k=0}^\infty$ and transfer function

$$H(z) = \sum_{k=0}^\infty h_k z^{-k} \tag{3.38}$$

is *stable* if bounded inputs u yield bounded outputs y.

RESULT B. A filter is stable if and only if its transfer function $H(z)$ converges absolutely on the unit circle, i.e.,

$$\sum_{k=0}^\infty |h_k| < \infty. \tag{3.39}$$

PROOF. Exercise 3.15.

As a corollary, if $H(z)$ is the transfer function of a stable filter, then the Laurent series

$$H(z) = \sum_{k=0}^{\infty} \frac{h_k}{z^k}$$

converges absolutely and uniformly for all $|z| \geq 1$ to a continuous function.

Example. The filter H with impulse response of powers $h_k = \alpha^k$ is clearly unstable for $|\alpha| > 1$ since not even the impulse response is bounded. If $|\alpha| = 1$, the response to the bounded constant signal $u_k = \bar{a}$ is unbounded by (3.31). If $|\alpha| < 1$, the response to any bounded input u_k is clearly bounded, as can be seen via the triangle inequality applied to (3.31). Result B obtains these three cases immediately since the series

$$\sum_{k=0}^{\infty} |\alpha|^k$$

converges exactly when $|\alpha| < 1$.

§3.5 Polar and Bode Plots

We have just seen in §3.3 that any causal signal h can be used as an impulse response to build a filter. But surely not all such filters will be useful. We need methods to judge the efficacy of various filters. There are two graphical methods, both built on the following fundamental insight.

THEOREM. Let F be a stable digital filter with real impulse response $h = \{h_k\}_{k=0}^{\infty}$ and transfer function $H(z)$. Then after transients have died away, the response to the sampled sinusoid $u_k = \sin \omega kT$ is again a sinusoid $y_k = r \sin(\omega kT + \phi)$ of the same frequency but with amplitude

$$r = |H(e^{i\omega T})| \qquad (3.40a)$$

and phase lead

$$\phi = \arg H(e^{i\omega T}). \qquad (3.40b)$$

PROOF. For $\xi = e^{i\omega T}$, find r and ϕ so that

$$H(\xi) = re^{i\phi}. \qquad (3.41)$$

Then the response y to the signal of powers of ξ, viz., $u = \{\xi^k\}_{k=0}^\infty$, is given by the convolution (3.31)

$$y_k = \sum_{j=0}^{k} \xi^{k-j} h_j = \xi^k \sum_{j=0}^{k} \xi^{-j} h_j$$

$$= \xi^k \sum_{j=0}^{\infty} \xi^{-j} h_j - \xi^k \sum_{j=k+1}^{\infty} \xi^{-j} h_j = \xi^k H(\xi) + o(1), \qquad (3.42)$$

where $o(1)$ denotes a sequence that is approaching 0 as $k \to \infty$. Thus the response to the sinusoid $\sin \omega k T = (\xi^k - \xi^{-k})/2i$ is by the above

$$\frac{\xi^k H(\xi) - \xi^{-k} H(\xi^{-1})}{2i} + o(1) = \frac{\xi^k r e^{i\phi} - \xi^{-k} r e^{-i\phi}}{2i} + o(1)$$

$$= r \frac{e^{i(\omega k T + \phi)} - e^{-i(\omega k T + \phi)}}{2i} + o(1) = r \sin(\omega k T + \phi) + o(1).$$

By this fundamental insight we may examine the performance of a given filter $y = h * u$ by examining its *polar plot*, i.e., the locus of all points in the complex plane $w = H(e^{i\omega T}) = r(\omega T)e^{i\phi(\omega T)}$ for all sinusoidal frequencies ω.

Example. The averaging filter $y_k = (u_k + u_{k-1})/2$ of (3.34) has transfer function $H(z) = (1 + z^{-1})/2$, so that

$$H(e^{i\omega T}) = (1/2)(1 + e^{-i\omega T})$$

$$= (1/2)(1 + \cos \omega T) - (1/2)i \sin \omega T, \qquad (3.43)$$

giving that

$$r = \frac{\sqrt{1 + \cos \omega T}}{\sqrt{2}}, \qquad \phi = -\arctan \frac{\sin \omega T}{1 + \cos \omega T}. \qquad (3.44)$$

Let us examine this filter when the sampling rate is 1000 samples per second.

```
% Routine 3.1 Polar plot
fS= 1000;                    % sampling rate in herz
T = 1/fS;                    % sampling period
f = 0:fS;                    % try frequencies 0 through fS
o = 2*pi*f;                  % radian frequency
```

```
w = exp(-i*o*T);        % w = 1/z
N = (1 + w)/2;          % transfer fn numerator
D = 1;                  % transfer fn denominator
H = N./D;               % transfer function H(z)
r = abs(H);             % filter gain
phi = angle(H);         % filter phase shift
polar(phi,r)            % polar plot
```

This polar plot is displayed in Figure 3.3.

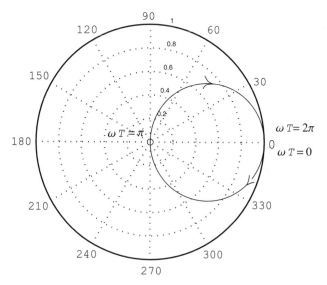

Figure 3.3. Polar plot of the filter (3.34).

Breaking a polar plot into two plots, where the gain $r = r(\omega T)$ and phase shift $\phi = \phi(\omega T)$ are separately graphed against the input frequency ω, yields the two *Bode* (pronounced "Bode-ee") *plots* of the filter $H(z)$.

Example. The Bode plots of the averaging filter $y_k = (x_k + x_{k-1})/2$ are shown in Figure 3.4. These plots were obtained by Routine 3.2 below. Often the gain $r = r(\omega T)$ is graphed in decibels of power $20 \log_{10} r$, as in Figure 3.5. Note the *lowpass* characteristic of this filter — higher frequencies are attenuated.

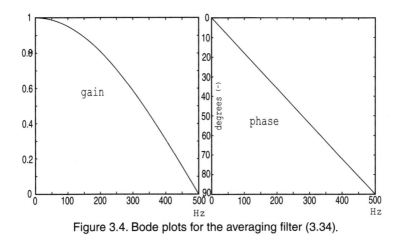

Figure 3.4. Bode plots for the averaging filter (3.34).

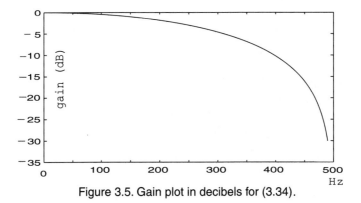

Figure 3.5. Gain plot in decibels for (3.34).

Example. The notch filter $y_k = u_k + u_{k-6}$ when the sampling rate is $f_S = 1/T = 720$ samples per second has the Bode plots shown in Figure 3.6. Notice the unwanted effects at frequencies other than 60 Hz.

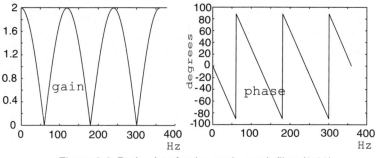

Figure 3.6. Bode plots for the crude notch filter (3.36).

Example. In contrast to the previous crude 60-Hz notch filter, consider the second-order filter

$$H(z) = \frac{(z - e^{120\pi i/720})(z - e^{-120\pi i/720})}{(z - 0.9e^{120\pi i/720})(z - 0.9e^{-120\pi i/720})}$$

$$= \cdots = \frac{1 - 2z^{-1}\cos(2\pi/12) + z^{-2}}{1 - 1.8z^{-1}\cos(2\pi/12) + 0.81z^{-2}}. \qquad (3.45)$$

The thought behind this design is to place a *zero* of $H(z)$ (a factor of the numerator) at the frequency we wish to notch out, then a stable pole (a factor of the denominator) nearby so that their ratio is nearly 1 at all other frequencies. The Bode plots shown in Figure 3.7 are generated by the following routine.

```
% Routine 3.2   Bode plots
fS= 720;                         % sampling rate in herz
T = 1/fS;                        % sampling period in seconds
f = 0:fS/2;                      % look at freqs.  up to fS/2
o = 2*pi*f;                      % radian frequencies
w = exp(-i*o*T);                 % w = 1/z
N = 1-2*w*cos(pi/6)+w.*w;        % transfer fn numerator
D = 1-1.8*w*cos(pi/6)+0.81*w.*w;% transfer fn denominator
H = N./D;                        % transfer function H(z)
r = abs(H);                      % filter gain
phi = angle(H);                  % filter phase shift (rad)
plot(f,r)                        % gain plot
% replace with plot(f,20*log10(r)) for gain decibel plot
% replace with plot(f,360*phi/(2*pi)) for phase plot
```

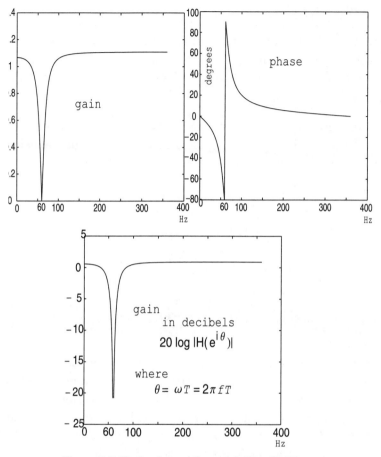

Figure 3.7. Bode plots of the notch filter (3.45).

Note the vast improvement in performance shown in Figure 3.7 over the filter of Figure 3.6. Frequencies away from 60 Hz are passed through the filter with little change in amplitude and phase. Demanding applications such as the EKG problem above may require the cascading of several second order filters of this type (see [Cadzow, p. 333]). Chebyshev, cyclotomic, and other special functions are used to generate filters since their zeros/poles are fortuitously located. There is, however, another dimension to filter design. In many applications the filter's *transient response* is an important consideration. Bode plots only represent the filter's performance after it has settled into periodic steady state (see Exercise 3.28).

§3.6 Aliasing

And now the bad news. Digital filters have a serious flaw inherited from the act of sampling itself: *It is impossible to distinguish between sampled signals of frequencies that are congruent modulo the sampling frequency.*

For example, if the sampling rate is $f_S = 1000$ Hz, then sinusoids of frequency 20 Hz, 1020 Hz, 2020 Hz, etc. are indistinguishable. The reason is simple: If

$$f \equiv f_0 \bmod f_S,$$

i.e., $f = f_0 + \mu f_S$, $\mu = 0, 1, 2, 3, \ldots$, then

$$\sin \omega kT = \sin 2\pi k fT = \sin \frac{2\pi k f}{f_S}$$

$$= \sin \frac{2\pi k (f_0 + \mu f_S)}{f_S} = \sin \frac{2\pi k f_0}{f_S} = \sin \omega_0 kT. \qquad (3.46)$$

You have most surely witnessed this phenomenon at the cinema, where the $f_S = 24$ Hz sampling camera records wagonwheels and propellers rotating slowly or in the retrograde direction. This phenomenon of *aliasing* is apparent in the conclusion of the theorem of §3.5: *Sampled filter response depends only on $\theta = \omega T$ modulo 2π.*

In fact, the situation is even worse. Since the response at the frequency ω is determined by

$$H(e^{i\omega T}) = re^{i\phi}, \qquad -\pi < \phi \leq \pi, \qquad (3.47)$$

and since $H(z)$ has real coefficients,

$$H(e^{-i\omega T}) = re^{-i\phi}. \qquad (3.48)$$

Thus *Bode gain plots are symmetric while phase plots are antisymmetric about half the sampling rate.* The gain symmetry is a serious defect while the phase plot antisymmetry (other than at zeros of H) is a harmless artifact of the branch of arg chosen in (3.47).

For example, the Bode plots of the averaging filter (3.34) when graphed out to the full sampling frequency $f_S = 1000$ Hz appear as in Figure 3.8.

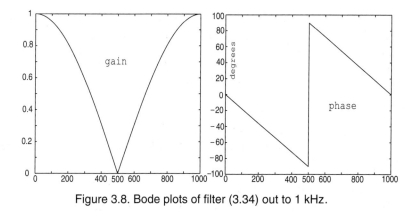

Figure 3.8. Bode plots of filter (3.34) out to 1 kHz.

In practice, hardware *antialiasing lowpass filters* are installed in the acquisition system before the analog-to-digital converter to roll-off the higher frequencies that would "fold over" onto the frequencies of interest.

Filter design is a profession in itself. There is a huge, ever-growing literature on digital filters. One place to begin is the superb 1973 text by Cadzow, or see the encyclopedic [Mitra and Kaiser]. The 1977 classic *Signal Analysis* by Papoulis is the standard theoretical reference.

§3.7 Closing the Loop

A computer is often used to monitor and steer an industrial process. A measurement of the process is sampled, then filtered and fed back to modify the parameters controlling the process — see Figure 3.9. The objective is to improve the robustness and accuracy of the process in the face of environmental perturbations (noise) and errors in the model.

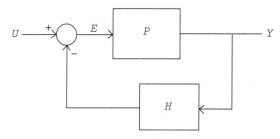

Figure 3.9. Closed-loop system.

Observation. In the closed loop of Figure 3.9, the frequency-domain output signal is on the one hand Y and on the other hand also the result of passing through the *plant* P the difference of the reference U less the filtered output HY; i.e.,

$$(U - HY)P = Y \tag{3.49}$$

so

$$Y = \frac{P}{1 + HP}\, U. \tag{3.50}$$

Example. Suppose our plant P is a not-so-ideal linear amplifier with gain β; i.e., $P(z) = \beta$. Suppose the gain β is too high or may suffer from instabilities, nonlinearities, or noise. To remediate, apply proportional negative feedback of gain $\alpha > 0$; i.e., $H(z) = \alpha$. Then, from (3.50),

$$\frac{Y}{U} = \frac{\beta}{1 + \alpha\beta}.$$

By adjusting the feedback gain α, we may lower the gain of the "compensated" amplifier in exchange for stability, linearity, and cleanliness. This is a universal practice in bipolar transistor amplifiers where a series resistor supplies the emitter.

Question. How can we be certain that the feedback loop has not introduced instabilities to the now closed-loop system?

Answer. Perform stability analysis using (3.39) or Exercise 3.16 on the closed-loop system (3.50). As long as the feedback transfer function H has no zeros at poles of the plant P, any instabilities of the original plant are canceled in the quotient (3.50); i.e., *feedback can be used to stabilize unstable plants,* provided that new unstable poles are not introduced by zeros of the denominator $1 + HP$.

Example. Consider the unstable plant $P(z) = (1 - 2z^{-1})^{-1}$ with the (unstable) pole at $z = 2$. Let us feed back an out-of-phase copy of the output Y using a linear amplifier $H(z) = \kappa$ of gain $\kappa = 3/2$ as in Figure 3.9 to obtain the closed-loop transfer function

$$\frac{P}{1 + HP} = \frac{(1 - 2z^{-1})^{-1}}{1 + (3/2)(1 - 2z^{-1})^{-1}} = \frac{2/5}{1 - (4/5)z^{-1}}. \tag{3.51}$$

Thus the unstable pole at $z = 2$ has been moved to the stable pole $z = 4/5$.

Problem. Design an automobile digital cruise control.

Solution. We obey the rule: *Design in the frequency domain, test in the time domain.* We sample vehicle speed Y at, say, $T = 0.1$, i.e., at $f_S = 10$ Hz (see Exercise 3.23). When the driver presses the cruise button, the present speed Y is recorded as the reference $Y_0 = \alpha + \alpha/z + \alpha/z^2 + \cdots$. Let us model the vehicle response with

$$Y = P\Theta - \gamma(L - L_0), \tag{3.52}$$

where θ is the angle of depression of the accelerator, where $U = \gamma(L - L_0)$ is the effect of the external load L (wind gusts, hills) less nominal load L_0 on accelerator response, and where

$$P(z) = \beta(1 + z^{-1} + z^{-2} + \cdots + z^{-n}), \quad \beta > 0; \tag{3.53}$$

i.e., the nominal speed response y to accelerator angle θ is proportional to the average of the last n readings of θ. With this model we attempt to capture the delay of the vehicle's response to changes in accelerator angle. (We derive a more physically based model in Chapter 10.) We feed back the error $E = Y - Y_0$ from set speed Y_0 through a filter/actuator H to displace accelerator angle θ from nominal θ_0, thus returning speed to the set speed (see Figure 3.10). A little algebra (Exercise 3.24) yields the relation between speed error $E = Y - Y_0$ and load perturbations $U = \gamma(L - L_0)$:

$$E = \frac{-1}{1 + PH}U. \tag{3.54}$$

Is simple proportional control $H(z) = \kappa > 0$ adequate to this task? Intuitively, a large value of feedback gain κ would make the transfer function E/U in (3.54) small, thereby washing out the effect of load change U on speed error E. On the other hand, large feedback gains κ might pump the accelerator wildly in response to varying load changes U. In fact, proportional feedback in (3.54) is stable for all $\kappa \geq 0$ (Exercise 3.25) and

$$\Theta - \Theta_0 = \frac{H}{1 + HP}U = \frac{\kappa}{1 + \kappa P}U \tag{3.55}$$

(Exercise 3.26), showing stable damped pedal movement in response to load changes. Proportional feedback $H(z) = \kappa$ appears viable.

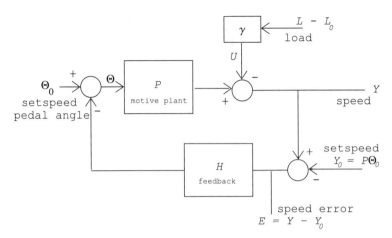

Figure 3.10. Cruise control system.

At the very least the closed loop system (3.54) must have an acceptable *step response* E to a unit *step* input $U = 1/(1 - z^{-1})$. The speed must return to approximately the set speed when climbing a hill of constant grade. By partial fractions, the step response from (3.54) is

$$E = \frac{-1}{1 + \kappa P}U = \frac{-1}{(1 + \kappa P(z))(1 - z^{-1})}$$

$$= \frac{\epsilon}{1 - z^{-1}} + \text{system modes,} \qquad (3.56)$$

where

$$\epsilon = \frac{-1}{1 + \kappa P(1)} = \frac{-1}{1 + \kappa \beta(n + 1)}, \qquad (3.57)$$

and where the *system modes* are the terms arising from the (stable) poles of the transfer function of the closed-loop system (3.54), i.e., decaying oscillations. Thus in the time domain, error from the set speed in response to a unit step in load is $e(t) = \epsilon + o(1)$ as $t \to \infty$, with ϵ presumedly small. To conclude this study we would simulate time-domain performance of this design using real-world estimates of the parameters and at various gains κ. We leave this as a major student project described in Exercise 3.29.

For further reading, start with the Schaum's Outline *Feedback and Control Systems* by DiStefano et al. One standard text is *Modern Control Systems* by Dorf. A more mathematical treatment is Sontag's *Mathematical Control Theory.*

§3.8 Why Decibels?

In acoustics, communication, seismology, and other fields it is common to express relative power levels in *decibels:*

$$\frac{p_2}{p_1} \text{(in decibels)} = 10\log_{10}\frac{p_2}{p_1} = 10\log_{10} p_2 - 10\log_{10} p_1. \qquad (3.58)$$

For instance, a drop of power output from 100 W to 10 W is -10 dB, a drop to 50 W is (approximately) -3 dB, a drop to 1 W is -20 dB, and so forth. The tiny radio-frequency (RF) power levels found in the front ends of receivers are often measured in *dBm,* i.e., in decibels relative to 1 milliwatt. For example, $10\,\mu\text{W} = -20$ dBm while 1W $= 30$ dBm.

Acoustic power level is commonly measured in decibels relative to the lowest audible *sound pressure level* (SPL) of 0.0002 dyn/cm^2. It is astonishing that this level is equivalent to detecting displacements of the eardrum less than one hundreths of the diameter of a hydrogen molecule! (See [Hartmann].) The brain/ear system is an extremely powerful signal processing system that can parse signal from noise at the quantum level [Bialek].

Recall from §3.5 that Bode gain plots are as a rule given in decibels of power referenced to a gain of 1. But why is the conversion shown in Figure 3.7 given as $20\log|H(e^{i\theta})|$? It is because most sensors transduce to voltage, so the variables of filter theory are in units of voltage. But power is the product of voltage v and current i, giving that the power p dissipated across a load of R ohms is $p = vi = v^2/R = i^2 R$. Thus power ratios in terms of voltage or current carry the multiplier 20 rather than 10.

But why decibels? There are two reasons: (1) to transform gain or loss calculations of cascaded systems to simple addition and subtraction, and (2) to comprehend ratios of vastly differing magnitudes.

Example. One lowpass filter rolls off -3 dB (down to half power) at 400 Hz while a second drops response -5 dB. Thus when these two filters are cascaded, the frequency 400 Hz is decreased -8 dB, i.e., $10^{-0.8} = 0.16$ of its original power level.

The *bandwidth* of a lowpass filter is the frequency of the -3 dB *half-power point*, i.e., where the filter passes only $1\sqrt{2}$ of the voltage that it passes at dc ($\omega = 0$).

Problem. We must set up a 5-GHz microwave data link from a remote site 12 km away. Both ends of the link employ small 10-dBi gain parabolic dishes. How much power must the transmitter run to yield a demodulatable signal at the receiver?

Solution. From their specifications we see that for successful de-modulation the receiver/modem requires a $v = 10$ μV signal across the $R = 50$-Ω Ohm receiver antenna connector, i.e., a signal of $30 + 10\log v^2/R = 30 + 10\log(10 \times 10^{-6})^2/50 = -87$ dBm. At both ends of the link there is 5 dB of transmission feedline loss.

The gain of each antenna is 10 dB over *isotropic*; i.e., in the preferred direction each watt of transmitted power is equivalent to 10 W distributed uniformly over a sphere centered at the parabolic dish. This power gain ratio $G = 10$ is determined by the rule

$$\text{gain } G = \frac{4\pi A}{\lambda^2}, \tag{3.59}$$

where A is the *effective aperture area* of the antenna and where λ is the radio wavelength [Kraus]. Our link will run at $f = 5$ GHz with wavelength $\lambda = c/f = 3 \times 10^8/(5 \times 10^9) = 6 \times 10^{-2}$ m = 6 cm.

Transmitted power is spread uniformly over an ever-growing sphere of area $4\pi r^2$. From (3.59) a unity-gain receiving antenna 12 km dis-tant intercepts only the portion of the power present on a spherical patch of area $A_0 = \lambda^2/4\pi$. Thus the *path loss over a distance r* is

$$10\log\frac{A_0}{4\pi r^2} = 10\log\frac{\lambda^2}{(4\pi r)^2} = 20\log\frac{\lambda}{r} - 20\log 4\pi$$

$$= 20\log 6 \times 10^{-5}/12 - 22 = -128 \text{ dB}. \tag{3.60}$$

For the link to succeed, *transmitted power + feedline losses + trans-mitter antenna gain + path loss + receiver antenna gain + feed-line losses must exceed the signal level required by the receiver*, i.e., $p - 5 + 10 - 128 + 10 - 5 > -87$, so the transmitter must supply an output of $p > 31$ dBm, i.e., $10^{3.1} = 1259$ mW = 1.3 W.

Exercises

3.1 Establish the z-transform (3.6). Find the z-transform of $x_k = \cos \omega kT$.

3.2 Find the z-transform of the signal $1, 2, 3, 1, 2, 3, 1, 2, 3, \ldots$.

3.3 Establish that the frequency-domain solution of (3.12) is (3.17).

3.4 Again solve recursion (3.12) but with arbitrary initial values x_{-1} and x_{-2}.

3.5 Solve the simple recursion of compound interest $x_k = (1 + i)x_{k-1}$ subject to $x_{-1} = 1/(1+i)$ using the z-transform method of §3.2.

3.6 The remaining balance x_k in a car loan account after making the kth monthly payment p is given by the recursion $x_k = (1+i)x_{k-1} - p$, where i is the monthly interest rate where x_0 is the original loan. Solve the recursion by methods of §3.2 and find the payment p that discharges the loan after N months.
Answer: $p = ix_0/[1 - (1+i)^{-N}]$.

3.7 Solve the recursion $6u_k = 5u_{k-1} - u_{k-2}$ when $u_0 = 1$ and $u_1 = -1$.

3.8 Solve $u_{k+1} - 2u_k + u_{k-1} = 0$ subject to $u_0 = 0$, $u_1 = 1$.

3.9 Determine that the total time T using (3.20) required to process a critical mass m of fissionable material using N identical reactors is $T = r\tau + (N-1)\tau$, where $r = m/V$ and where V is the capacity of each reactor.

3.10 (**Project**) Look again at the reactor equation (3.20). Given fixed reaction rates $\alpha > \beta$ determined by the compounds themselves, we might obtain higher purity at the Nth step by using various longer but decreasing holding times τ_k at the kth reactor of the cascade. On the other hand, we may be under the deadline $r\tau_1 + \tau_2 + \cdots + \tau_N = T$. Assign values to $\alpha > \beta$ and fix the number of reactors N, the number of fractions r that must be processed, the beginning and ending concentrations $x_0 < x_N$, and a deadline T. Design a Monte Carlo routine that will search for the optimal sequence of holding times $\tau_1 \geq \tau_2 \geq \cdots \geq \tau_N$.

3.11 Sketch a plan for becoming rich beyond the dreams of avarice should a design for noncausal filters come into your possession.

3.12 Show that the filter $F = 1 + z^{-6}$ of §3.3 where the sampling period is $T = 1/720$ s will notch out not only 60 Hz but also any odd multiple of 60 Hz. What does it do to even multiples of 60 Hz?

Hint: $u_k = \sin(\mu 2\pi k/12)$ where $\mu = 1, 2, 3, \ldots$.

3.13 Find the transfer function of the matched filter (3.37).

3.14 (**Project**) Using the Gaussian noise generator `randn` of MATLAB, simulate the effect of a matched filter of length $m = 10$ periods on a noisy sinusoidal $u_k = 0.2 \sin 2\pi k/12 +$ `randn`. Graph the signals before and after processing. Experiment with more periods m and other sampling rates. If the hardware is available, listen to the improvement at various signal frequencies f_0 between 400 and 2000 Hz via the MATLAB command `wavwrite`.

3.15* Show that if $H(1)$ converges absolutely, the filter $H(z)$ is stable. Attempt the difficult converse.

Remark: Although an elementary proof of the converse is very difficult, it follows easily from the closed graph theorem of functional analysis applied to l^∞ (see the proof of Theorem B in §10.5). The converse also becomes easy with the additional assumption that $H(z)$ is rational (see Exercise 3.16).

3.16 Prove that a filter with a proper rational transfer function

$$H(z) = \frac{b_0 + b_1 z^{-1} + \cdots + b_m z^{-m}}{a_0 + a_1 z^{-1} + \cdots + a_n z^{-n}}, \quad \text{with } m \le n, \text{ and } a_0 \ne 0$$

(in lowest terms) is stable if and only if all its poles lie within the open unit disk.

Hint: Think partial fractions.

3.17 Which proper rational transfer functions belong to FIR filters? [A filter has a *finite impulse response* (FIR) $h = \{h_k\}_{k=0}^\infty$ if $h_k = 0$ for all large k.]

3.18 Experiment with Routine 3.1. Construct the polar plots of the notch filters (3.36) and (3.45).

3.19 Obtain the Bode plots of the notch filter obtained by cascading two identical notch filters (3.45). You need add but one line to Routine 3.2.

3.20 Referring to Figure 3.10, show that the unstable plant $P(z) = 1/(1-\alpha z^{-1})$, with $|\alpha| > 1$ is stabilized by proportional feedback $H(z) = \kappa$ for any positive gain $\kappa > |\alpha| - 1$.

3.21 Test the transfer function $H(z) = 1/(1 + 5z^{-1} + 7z^{-2})$ for stability. More generally, formulate a stability criterion for second-order plants

$$H(z) = \frac{d + ez^{-1} + fz^{-2}}{a + bz^{-1} + cz^{-2}}, \qquad a \neq 0, b, c \text{ real.}$$

3.22 Test the impulse response $h = \{1/[(k+2)\ln(k+2)]\}_{k=0}^{\infty}$ for stability.

3.23 (**Project**) How would you measure automobile speed for an input to a cruise control system? What transducer/sensor would you employ? Where would you tap off the measurement? What sampling period T would be sufficient? After writing down your best guess, find out how it is actually done.

3.24 Deduce (3.54) from Figure 3.10.

3.25*Apply Exercise 3.16 to show that proportional cruise control $H(z) = \kappa$ in (3.54) yields a stable system for all $\kappa \geq 0$.

3.26 Derive (3.55) from Figure 3.10.

3.27 Prove that if a stable transfer function $H(z)$ has impulse response $h = \{h_k\}_{k=0}^{\infty}$, then

$$h_k = \frac{1}{2\pi} \int_0^{2\pi} H(e^{i\theta}) e^{ik\theta} \, d\theta.$$

3.28 Experiment with the 60-Hz notch filter (3.45). Move the stable poles at $z = 0.9e^{\pm i\pi/6}$ (in the denominator) closer to the unit circle by replacing 0.9 by 0.999. Note how the Bode plot improves. However, the transient response seriously degrades: Compare the first several periods of the initial response of both filters to a 60-Hz sinusoid.

3.29 (Project) Simulate the performance of the proportional feed-back cruise control system designed in §3.7. Estimate β and n of the motive plant $P(z)$ to accurately model the pedal response of a known vehicle. Estimate conceivable amplitudes of a step change in load U. Write a routine that computes the first N terms of the impulse response of the transfer function (3.54), then the approximate step response. Experiment with various gains κ. How sensible is a linear model since wind resistance increases more or less as the square of speed? Finally, decide whether the pedal transfer function $P(z)$ in (3.53) is a reasonable model and if proportional feedback $H(z) = \kappa$ yields acceptable performance.

3.30 Argue that a good model for integration

$$y(t) = \int_0^t u(\tau)\, d\tau$$

sampled at intervals of time T is

$$y_k = y_{k-1} + \frac{T(u_k + u_{k-1})}{2}$$

i.e.,

$$Y = \frac{T}{2} \frac{1 + z^{-1}}{1 - z^{-1}} U.$$

3.31 From inverting the transfer function for integration in Exercise 3.30 comes *Tustin's method* for solving ordinary differential equations: Replace differentiation with respect to time by

$$\frac{2}{T} \frac{1 - z^{-1}}{1 + z^{-1}}.$$

For example, the solution y to the first-order ordinary differential equation

$$\dot{y} + 2y = 1, \quad y(0) = 0$$

when sampled at $T = 0.1$ s satisfies approximately

$$20 \frac{1 - z^{-1}}{1 + z^{-1}} Y + 2Y = \frac{1}{1 - z^{-1}}.$$

Solve this recursion and compare your solution to the actual solution.

3.32*Solve via Tustin's method the ODE $\ddot{y} - \dot{y} + y = u$ subject to $y(0) = \dot{y}(0) = 0$. Compare the continuous to the discrete solution for the unit step $[u(t) = 1, \ t \geq 0$, zero otherwise] when $T = 0.001$.

Answer:

$$y_k = \frac{1}{4 - 2T + T^2}\left[-(2T^2 - 8)y_{k-1} - (4 + 2T + T^2)y_{k-2}\right.$$

$$\left. + T^2 u_k + 2T^2 u_{k-1} + T^2 u_{k-2}\right].$$

3.33*Prove the *small gain theorem:* For $P(z)$ stable and meromorphic,

$$G(z) = \frac{1}{1 + kP(z)}$$

is stable for all small $|k|$.

3.34 Display a stable transfer function $H(z)$ that is neither a rational function nor a FIR.

3.35 Why is it important that a bandpass filter have *constant group delay* throughout its passband, i.e., that the phase lead/lag ϕ in (3.40b) be a linear function of frequency ω?

Hint: Simultaneous zero crossings of signals in the passband must be preserved to prevent *phase distortion.*

3.36 A 1-W data transmitter is sitting on the (floating?) ice of Europa, 630 million kilometers distant. It is beaming data at a carrier frequency of $f = 3$ GHz via a 20-dBi-gain parabolic dish aimed back at Earth. Estimate the physical size of an Earth-based antenna required to capture -172 dBm of this distant signal.

Answer: A 316-m^2 antenna.

3.37 (**Project**) Experiment with image processing. Using the MATLAB commands `colormap` and `image`, construct a rudementary gray-scale 20 × 20 pixel image that is a visual representation of the entries of a 20 × 20 matrix. Using `randn`, superimpose Gaussian noise onto a clean image. Clean up the image by using a weighted average of nearby gray-scale values. Sharpen edges with numerical differentiation. Read and report on other processing methods. Make a movie.

3.38 Find the kth Fibonacci number x_k where $x_{k+2} = x_{k+1} + x_k$ with $x_0 = x_1 = 1$.

3.39**A rich complicated signal $x = \{x_k\}_{k=0}^{\infty}$ may often be given at least approximately as the impulse response of a rational function $B(z)/A(z)$ of low order. Then in order to conserve communication bandwidth, one transmits the (small number of) coefficients of B/A rather than the lengthy signal x. Which signals x can be approximated to any desired accuracy (and in what sense) by the impulse response of second-order filters

$$F(z) = \frac{b_0 + b_1 z^{-1} + b_2 z^{-2}}{a_0 + a_1 z^{-1} + a_2 z^{-2}}?$$

3.40 (Project) Experimentally design a *stopband* filter that suppresses frequencies between 60 and 100 Hz by cascading three filters as in (3.45).

3.41 (Project) In contrast to Exercise 3.14, filter a noisy sinusoid with a peak filter $F(z) = 1 - H(z)/|H(\omega_0)|$, where $H(z)$ is a notch filter of type (3.45) centered at the sinusoid's carrier frequency ω_0. Demonstrate the improved signal-to-noise ratio graphically, and if possible aurally, by means of a sound card. How effective is the simple averaging filter (3.34)?

3.42 Find the z-transform of $x_k = \sinh \omega k T$.

3.43 Find the z-transform of $x_k = kT$.

3.44 Find the z-transform of $x_k = 1/(k+1)T$.

3.45 Estimate the rotational rate in rpm of the wagonwheels seen in Western movies in two ways — first from the apparent aliasing of the 24-Hz camera sampling rate, then by wheel size and wagon speed.

Chapter 4

The Discrete Fourier Transform

Spectral analysis of sampled data is accomplished by the discrete Fourier transform (DFT).

How is real-time spectral analysis performed on incoming data? It is done via the discrete Fourier transform. We study the properties of this transform with its ability to transform convolutions to products. This viewpoint will give us insight into FIR filters. We shall learn how in practice the DFT is done via the fast Fourier transform. We walk through the synthesis of an instrument like the oboe. Finally, we practice image encryption and enhancement.

§4.1 Real-time Processing

Let us take a more practical view than in Chapter 3. We wish to perform spectral analysis and filtering on data acquired from a process as in Figure 3.1, and we wish to do this in *real time* or *on-line*, as the process is occurring — not *off-line*, after the fact. We are limited by our computer's speed and memory — by how many samples can be retained and processed as a unit quickly enough to keep pace with incoming data (after a short but acceptable delay).

Assume that we can keep up with events when samples are arriving at some rate using retained data length n. Set

$$\xi = e^{2\pi/n} = \cos\frac{2\pi}{n} + i\sin\frac{2\pi}{n}, \qquad (4.1)$$

the primitive nth root of unity of least positive angle. (Many prefer the notation W or w rather than ξ, presumedly from the German word *Wurzel* for "root.")

The *frame* of accumulated data

$$x = (x_0, x_1, x_2, \ldots, x_{n-1}) \qquad (4.2a)$$

is transformed into the *frequency-domain* object

$$\hat{x} = (\hat{x}_0, \hat{x}_1, \hat{x}_2, \ldots, \hat{x}_{n-1}) \qquad (4.2b)$$

by the *discrete Fourier transform* (DFT)

$$x \mapsto \hat{x}, \qquad (4.3a)$$

where

$$\hat{x}_k = \sum_{j=0}^{n-1} x_j \xi^{-jk}, \qquad (4.3b)$$

i.e.,

$$
\begin{vmatrix}
\hat{x}_0 \\
\hat{x}_1 \\
\hat{x}_2 \\
\hat{x}_3 \\
\vdots \\
\hat{x}_{n-1}
\end{vmatrix}
=
\begin{pmatrix}
1 & 1 & 1 & 1 & \cdots & 1 \\
1 & \xi^{-1} & \xi^{-2} & \xi^{-3} & \cdots & \xi^{-(n-1)} \\
1 & \xi^{-2} & \xi^{-4} & \xi^{-6} & \cdots & \xi^{-2(n-1)} \\
1 & \xi^{-3} & \xi^{-6} & \xi^{-9} & \cdots & \xi^{-3(n-1)} \\
\vdots & & & & & \vdots \\
1 & \xi^{-(n-1)} & \xi^{-2(n-1)} & \xi^{-3(n-1)} & \cdots & \xi^{-(n-1)^2}
\end{pmatrix}
\begin{vmatrix}
x_0 \\
x_1 \\
x_2 \\
x_3 \\
\vdots \\
x_{n-1}
\end{vmatrix}
.
$$

Thus thinking of the frames x and $\hat{x}$ as column vectors,

$$\hat{x} = Fx. \qquad (4.4)$$

But for any integer a, there is the famous relation

$$\sum_{k=0}^{n-1} \xi^{ak} = \begin{cases} n & \text{if } \xi^a = 1 \\ 0 & \text{otherwise} \end{cases} \qquad (4.5)$$

(Exercise 4.1), giving that

$$F\bar{F} = nI \qquad (4.6)$$

(Exercise 4.2), where the overbar is complex conjugation. Therefore, from (4.4) and (4.6) comes the *inverse discrete Fourier transform*

$$x = \frac{1}{n}\bar{F}\hat{x}, \qquad (4.7)$$

i.e.,

$$x_k = \frac{1}{n} \sum_{j=0}^{n-1} \hat{x}_j \zeta^{jk}. \tag{4.8}$$

The DFT is actually an older tool than its cousins, the Fourier transform and Fourier series, according to the historical notes in Chapter 1 of [Briggs and Henson]. One can choose to view the DFT as merely a clever computational device — when coupled with the *fast Fourier transform* (FFT), the DFT makes certain real-time computations possible at throughputs much higher than if calculations were limited to the time domain [Strang]. Oddly enough, it is faster to transform a signal into the frequency domain, filter, then transform back, than to do the filtering (convolutions) in the time domain.

A more theoretical view is that the DFT is an approximation (Riemann sum) of the important Fourier coefficients that give the frequency content of periodic signals — see Chapter 11 and Exercise 4.8. Or we will see in Chapter 6 that the DFT is the best spectral fit in energy to a signal x.

§4.2 Properties of the DFT

Because of (4.4) and (4.7), the DFT

$$x \mapsto \hat{x}$$

is clearly a linear bijective map from the complex vector space $\mathbf{C}^n$ onto itself. Its importance lies in the transformation of convolution to product.

The *circular convolution* of two n-tuples x and y,

$$x * y = z,$$

is again an n-tuple z given by the rule

$$z_k = \sum_{j=0}^{n-1} x_j y_{k-j}, \tag{4.9}$$

where subscripts are computed modulo n.

Example. Given that $x = (1, 1, 2)$ and $y = (-1, 3, 4)$, their circular convolution is $z = x * y = (9, 10, 5)$ since

$$z_0 = 1 \cdot (-1) + 1 \cdot 4 + 2 \cdot 3$$
$$z_1 = 1 \cdot 3 + 1 \cdot (-1) + 2 \cdot 4$$
$$z_2 = 1 \cdot 4 + 1 \cdot 3 + 2 \cdot (-1).$$

REMARK. This rather odd convolution can be recognized as the ordinary discrete convolution filtering action

$$y_k = \sum_{j=0}^{k} h_j u_{k-j}$$

of Chapter 3 in the special case that the input u is periodic every n samples, and that h is a finite impulse response (FIR) filter of length n (see Figure 4.1).

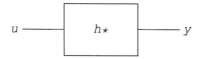

Figure 4.1. Filtering by convolution.

Our new and more practical game is to assume the incoming data is temporarily periodic, to filter as if it were periodic, to obtain its spectrum as if it were periodic, and to output the result as a processed n-tuple frame. Meanwhile, during these calculations, the acquisition system is grabbing and storing a new frame of readings for subsequent processing.

RESULT A. The transform of convolution is coordinate-wise product. The transform of coordinate-wise product is $1/n$ times convolution. In symbols,

$$x * y \mapsto \hat{x} \cdot \hat{y} \tag{4.10a}$$

and

$$x \cdot y \mapsto \frac{1}{n} \hat{x} * \hat{y}. \tag{4.10b}$$

PROOF. Let $z = x * y$. Then

$$\hat{z}_k = \sum_{j=0}^{n-1} z_j \xi^{-jk} = \sum_{j=0}^{n-1} (x * y)_j \xi^{-jk} = \sum_{j=0}^{n-1} \left(\sum_{m=0}^{n-1} x_m y_{j-m} \right) \xi^{-jk}$$

$$= \sum_{m=0}^{n-1} x_m \sum_{j=0}^{n-1} y_{j-m} \xi^{-jk} = \sum_{m=0}^{n-1} x_m \xi^{-km} \sum_{j=0}^{n-1} y_{j-m} \xi^{-(j-m)k}$$

$$= \Big(\sum_{m=0}^{n-1} x_m \xi^{-km} \Big) \Big(\sum_{j=0}^{n-1} y_j \xi^{-jk} \Big) = \hat{x}_k \cdot \hat{y}_k,$$

giving (4.10a).

As for (4.10b), start instead from the right-hand side: If $\hat{z} = \hat{x} * \hat{y}$, by (4.5),

$$\hat{z}_k = \sum_{j=0}^{n-1} \hat{x}_j \hat{y}_{k-j} = \sum_{j=0}^{n-1} \Big(\sum_{p=0}^{n-1} x_p \xi^{-pj} \Big) \Big(\sum_{q=0}^{n-1} y_q \xi^{-q(k-j)} \Big)$$

$$= \sum_{p,q=0}^{n-1} x_p y_q \xi^{-qk} \sum_{j=0}^{n-1} \xi^{j(q-p)} = n \sum_{p=0}^{n-1} x_p y_p \xi^{-pk},$$

which is n times the transform of $x \cdot y$.

COROLLARY. Filtering a periodic input $u = (u_0, \dots, u_{n-1})$ by circular convolving by the finite impulse response, $h = (h_0, \dots, h_{n-1})$,

$$y = h * u$$

becomes pointwise product in the frequency domain; i.e.,

$$\hat{y} = \hat{h} \cdot \hat{u}.$$

§4.3 Filter Design

All the results of §3.5 on Bode plots and frequency-domain response remain valid here for the special case of this chapter — for signals u that are periodic every n samples and FIR filters of length n. As guaranteed by the theorem of §3.5, the periodic steady-state response at frequency ω of a filter with impulse response $h = \{h_k\}_{k=0}^{n-1}$ is given by the transfer function $H(z)$ evaluated at $z = e^{i\theta}$, $\theta = \omega T$, yielding the gain

$$r = |H(e^{i\theta})| \tag{4.11a}$$

and phase lead

$$\phi = \arg H(e^{i\theta}). \tag{4.11b}$$

But note that

$$H(\xi^k) = \hat{h}_k, \tag{4.12}$$

so *the discrete Fourier transform of the impulse response h samples the gain r and phase lead ϕ of the filter at frequencies corresponding to vertices of the regular n-gon inscribed within the unit circle.*

Problem. When sampling at $f_S = 720$ Hz, design a filter to notch out the extraneous 60-Hz signal induced by room wiring.

Solution. We have already solved this problem twice — see (3.36) and (3.35). Here is a third attempt.

Take frame length $n = f_S = 720$. We want the gain at 60 Hz to be zero, i.e., $|\hat{h}_{60}| = 0$, yet gains at other frequencies to be 1, i.e., $|\hat{h}_k| = 1$ for $k \neq 60$. This is too much to hope, since if you recall from (3.48), the gain response of a filter is symmetric about half the sampling frequency. Thus we must also assign $|\hat{h}_{720-60}| = |\hat{h}_{660}| = 0$. This is a special case of the general description of any real FIR.

RESULT B. A FIR filter of frame length n with impulse response h with DFT $\hat{h}_k = r_k e^{i\phi_k}$ is real if and only if

$$r_{n-k} = r_k \quad \text{and} \quad \phi_{n-k} \equiv -\phi_k \bmod 2\pi. \tag{4.13}$$

PROOF. Exercise 4.4.

In view of this result we must be careful with our choice of the argument ϕ_k of each h_k in our 60-Hz notch filter. The simplest choice would be to take phase 0 for each term, i.e.,

$$\hat{h}_k = \begin{cases} 0 & \text{if } k = 60 \text{ or } 660 \\ 1 & \text{otherwise when } 0 \leq k \leq 721. \end{cases} \tag{4.14}$$

This results in a *terrible* filter with unacceptable time-domain transient response (Exercise 4.6). See Exercise 4.5 for a guide to remediation.

But conceptually, FIR filter design seems straightforward:

Step 1. Draw the Bode gain and phase plot of a desired filter as in Figure 4.2.

Step 2. Partition the abscissa $0 \leq \theta \leq 2\pi$ into n equal parts $0 < \theta_1 < \theta_2 < \cdots < \theta_{n-1} < 2\pi$.

Step 3. Choose $\hat{h}_k = r_k e^{i\phi_k}$ so that r_k is the gain at θ_k and ϕ_k the phase, as in Figure 4.3.

Step 4. Transform $\hat{h}$ back to obtain the impulse response h of your desired filter.

Better yet, do the real-time filtering by transforming the input u to the frequency domain, perform coordinate-wise multiplication with $\hat{h}$, then transform the result $\hat{y} = \hat{h} \cdot \hat{u}$ back to the time domain — the transforms in both directions done by an FFT.

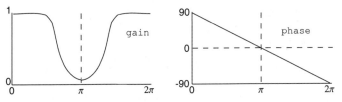

Figure 4.2. Begin by drawing the desired Bode plots.

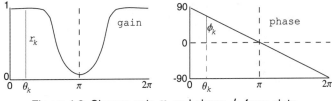

Figure 4.3. Choose gain r_k and phase ϕ_k from plots.

This simpleminded approach to design will seldom yield well behaved filters in the time domain, but does provide a conceptual hook on which to hang your intuition — my only goal for this section. Consult design books such as [Mitra and Kaiser] or [Rorabaugh] for careful design techniques.

§4.4 The Fast Fourier Transform

Performing the DFT

$$\hat{x}_k = \sum_{j=0}^{n-1} x_j \xi^{-jk} \tag{4.15}$$

is expensive since it calls on transcendental central processing unit support for $\xi^{-jk} = \cos 2\pi jk/n + i \sin 2\pi jk/n$ and demands n^2 multiplications be performed. The values of ξ^{-jk} can be computed offline and stored for the calling, but the n^2 multiplications remain. Astonishingly, if $n = 2^\alpha$, this transform can be done using only approximately n multiplications. Although well known to Gauss, this *fast Fourier transform* is associated with the contemporary names Cooley and Tukey (see the historical notes in Chapter 10 of [Briggs and Henson]).

The FFT is based on the following simple observation for even frame lengths $n = 2m$:

$$\hat{x}_k = x_0 + x_1 \xi^{-k} + x_2 \xi^{-2k} + \cdots + x_{n-1} \xi^{-(n-1)k}$$

$$= x_0 + x_2(\xi^2)^{-k} + x_4(\xi^2)^{-2k} + x_6(\xi^2)^{-3k} + \cdots + x_{2(m-1)}(\xi^2)^{-(m-1)k}$$

$$+ \xi^{-k} \cdot (x_1 + x_3(\xi^2)^{-k} + x_5(\xi^2)^{-2k} + x_7(\xi^2)^{-3k} + \cdots + x_{2m-1}(\xi^2)^{-(m-1)k}). \tag{4.16}$$

That is,

$$\hat{x}_k = \hat{y}_k + \xi^{-k} \hat{z}_k, \tag{4.17}$$

where

$$\hat{y}_k = y_0 + y_1 \zeta^{-k} + y_2 \zeta^{-2k} + \cdots + y_{m-1} \zeta^{-(m-1)k} \tag{4.18a}$$

and

$$\hat{z}_k = z_0 + z_1 \zeta^{-k} + z_2 \zeta^{-2k} + \cdots + z_{m-1} \zeta^{-(m-1)k}, \tag{4.18b}$$

where

$$\zeta = \xi^2, \tag{4.18c}$$

the primitive mth root of unity of least positive argument. That is, *a DFT of frame length $n = 2m$ is the weighted sum of two DFTs of half the frame length m.*

In fact, although (4.17) appears to require a multiplication for each of the n values $k = 0, 1, 2, \ldots, n - 1$, since $\xi^{n/2} = -1$, we see instead that

$$\hat{x}_k = \hat{y}_k + \xi^{-k} \hat{z}_k \tag{4.19a}$$

but

$$\hat{x}_{m+k} = \hat{y}_k - \xi^{-k}\hat{z}_k, \tag{4.19b}$$

for $k = 0, 1, \ldots, m-1$.

We can think of (4.19) as a matrix factorization of the nth Fourier matrix $F_n = (\xi^{-jk})$ of (4.4):

$$F_n = B_n \begin{pmatrix} F_m & 0 \\ 0 & F_m \end{pmatrix} P_n \tag{4.20}$$

where B_n is the *butterfly* matrix

$$B_n = \begin{pmatrix} I_m & \Xi_m \\ I_m & -\Xi_m \end{pmatrix}, \tag{4.21a}$$

with I_m the $m \times m$ identity matrix and Ξ_m the $m \times m$ diagonal matrix $\Xi_m = \mathrm{diag}(1, \xi_n^{-1}, \xi_n^{-2}, \ldots, \xi_n^{-(m-1)})$, (where ξ_n denotes the canonical primitive nth root of unity) and where P_n is a permutation matrix of 0's and 1's that reorders the components of the frame into the even- then odd-subscripted components:

$$(x_0, x_1, x_2, x_3, \ldots, x_{n-1}) P_n^T$$

$$= (x_0, x_2, x_4, \ldots, x_{n-2}; x_1, x_3, x_5, \ldots, x_{n-1}). \tag{4.21b}$$

So if $n = 2^\alpha$, we can continue to decompose the DFT into DFTs of lower and lower frame lengths by successively factoring the Fourier matrix. But note that

$$B_2 = F_2 = \begin{pmatrix} 1 & 1 \\ 1 & -1 \end{pmatrix} \tag{4.22}$$

and $F_1 = 1$ where the successive factorization finally breaks off. Two successive factorizations would look like

$$F_n = B_n \begin{pmatrix} B_{n/2} & 0 \\ 0 & B_{n/2} \end{pmatrix}$$

$$\cdot \begin{pmatrix} F_{n/4} & 0 & 0 & 0 \\ 0 & F_{n/4} & 0 & 0 \\ 0 & 0 & F_{n/4} & 0 \\ 0 & 0 & 0 & F_{n/4} \end{pmatrix} \begin{pmatrix} P_{n/2} & 0 \\ 0 & P_{n/2} \end{pmatrix} P_n,$$

and so forth.

In any implementation of this algorithm, once the frame length $n = 2^\alpha$ is chosen, the permutation matrices are multiplied out into one permutation matrix Q off-line to form a look-up table for reordering the components x_k. Also, the complex matrices Ξ_k are tabulated off-line and stored. The FFT algorithm obtains the DFT $\hat{x}$ of the frame x by operating right to left in the iterative process

$$\hat{x} = B_n \begin{pmatrix} B_{n/2} & 0 \\ 0 & B_{n/2} \end{pmatrix} \cdots$$

$$\cdot \begin{pmatrix} B_4 & 0 & . & . & 0 \\ 0 & B_4 & 0 & . & 0 \\ & & . & & \\ 0 & & & & B_4 \end{pmatrix} \begin{pmatrix} B_2 & 0 & . & . & . & 0 \\ 0 & B_2 & 0 & . & . & 0 \\ & & . & & & \\ & & & . & & \\ 0 & & & & & B_2 \end{pmatrix} Qx.$$

Problem. What makes the oboe so distinctive an instrument?

Solution. Suppose we have digitally sampled 1 second of an oboe playing A 440 Hz at a sampling rate of $f_S = 8192 = 2^{13}$ samples per second. Let us assume that the sound content of an oboe is predominately the fundamental tone, with an additional second harmonic, a slightly out-of-tune third harmonic, and wind noise. I invite you in Exercise 4.7 to obtain the actual facts. Using MATLAB we construct a practice sampled data set as follows:

```
fS = 8192;                    % set sampling rate
t = 0:fS-1;                   % sampling vector
t = t/fS;                     % sampling times
w = 2*pi*440;                 % radian A 440 Hz
x = sin(w*t);                 % fundamental
x = x + .2*sin(2*w*t);        % first overtone
x = x + .1*sin(3.01*w*t)      % second overtone
x = x + .4*randn(1,fS);       % wind noise
```

Let us perform a DFT on this data via the FFT of MATLAB:

```
% Routine 4.1 Spectral analysis
n= 8192;                      % set FFT frame length
m = 400;                      % set viewing window min.  herz
```

```
M = 1400;              % set viewing window max.  herz
xhat = fft(x,n);       % perform FFT
xhat = xhat/n;         % normalize (see Exercise 4.8)
f = 0:n-1;             % abscissa in herz
f = f(m:M);            % abscissa viewing window
yhat = xhat(m:M);      % ordinate viewing window
plot(f,abs(yhat))      % the spectrum
```

The computed normalized spectrum is shown in Figure 4.4.

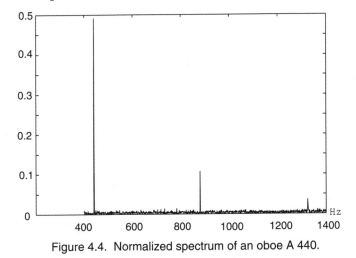

Figure 4.4. Normalized spectrum of an oboe A 440.

By (4.12), the remaining half of the spectral content is found at the conjugate frequencies as seen in Figure 4.5.

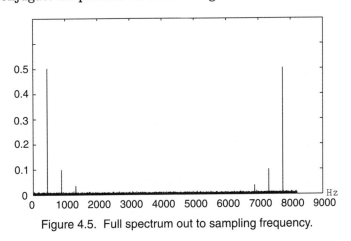

Figure 4.5. Full spectrum out to sampling frequency.

WARNING. Figures 4.4 and 4.5 show the magnitudes $|\hat{x}_k|$ of the transform coefficients normalized by frame length n, not the *power spectral density* (see [Papoulis]). The power spectral density is $(\pi/2)|\hat{x}_k/n|^2$ in watts per hertz.

REMARK. In practice the sampling rate f_S is much larger than the length n of the DFT frame, so that there are f_S/n frames being processed per second. Thus the abscissa of a spectral display must be adjusted for units by this same ratio $r = f_S/n$ (see Exercise 4.9).

§4.5 Image Processing

An interesting application of the DFT is to image processing. Think of an image as composed of $m \times n$ pixels with each pixel (i, j) assigned some integer intensity $0 \le x_{ij} \le W$ along a *gray scale* from 0 (black) to W (white). For instance, arrays of size $m = n = 128$ and gray-scale resolution $W = 255$ give recognizable images of a human face. So for us here, an *image* X is a $m \times n$ matrix (x_{ij}) of integers between 0 and W.

Images are enhanced by performing some filter H on the image X to create a new image Y. For example, contrast is enhanced by lowering low-intensity pixels to an even lower intensity, and increasing the intensity of the higher. One such is the *gamma correction*,

$$y_{ij} = [W \cdot (\frac{x_{ij}}{W})^\gamma], \tag{4.23}$$

while another is *thresholding*,

$$y_{ij} = \begin{cases} W & \text{if } x_{ij} \ge k \\ 0 & \text{otherwise.} \end{cases} \tag{4.24}$$

There are many variations on this idea of pixel-by-pixel enhancement, with names like *intensity shift, intensity multiply, logarithm, negate, highlight, squaring*, etc. (see [Batchelor and Waltz] or [Jain et al.]).

But the more powerful image processing routines use not only the intensity at the pixel to be modified, but the intensities of surrounding pixels. Linear such routines — weighted averages of nearby intensities — are in fact two-dimensional circular convolutions $Y = H * X$:

$$y_{ij} = \sum_{p,q=1}^{m,n} h_{p,q} x_{i-p,j-q} \tag{4.25}$$

with subscripts computed modulo m and n, respectively. Typically, H is sparse of the form

$$H = \begin{pmatrix} K_{r\times s} & 0 & 0 \\ 0 & 0 & 0 \\ 0 & 0 & L_{r\times s} \end{pmatrix} \tag{4.26}$$

with $r, s \ll m, n$. Such filters H will cause *edge effects* at the borders of the image since the wraparound convolution (4.25) uses intensities from near the opposite border. This untoward effect is corrected by cropping the filtered image.

Images X transmitted through a channel with known linear *blurring* B yielding the blurred images $Y = B * X$ can be *deblurred* by inverting the convolution. As you might have guessed, this is done by the DFT.

The *discrete Fourier transform* of a $m \times n$ matrix $X = (x_{ij})$ is the $m \times n$ matrix $\hat{X} = (\hat{x}_{ij})$, where

$$\hat{x}_{ij} = \sum_{p,q=1}^{m,n} x_{pq} \zeta_m^{-pi} \zeta_n^{-qj}, \tag{4.27a}$$

where ζ_k denotes the primitive kth root of unity of least positive angle. The inverse transform is

$$x_{ij} = (mn)^{-1} \sum_{p,q=1}^{m,n} \hat{x}_{pq} \zeta_m^{pi} \zeta_n^{qj} \tag{4.27b}$$

(Exercise 4.10).

Not unexpectedly, the DFT $X \to \hat{X}$ carries convolution to entry-wise product, and entry-wise product to $(mn)^{-1}$ times convolution:

$$X * Y \to \hat{X} \cdot \hat{Y} \tag{4.28a}$$

$$X \cdot Y \to (mn)^{-1} \hat{X} * \hat{Y} \tag{4.28b}$$

(Exercise 4.11).

Thus *to deblur, transform the blurred image $Y = B * X$ to its frequency-domain alter ego $\hat{Y}$, divide each entry by the corresponding entry of blurring $\hat{B}$, then transform back*, i.e.,

$$x_{ij} = (mn)^{-1} \sum_{p,q=1}^{m,n} \frac{\hat{y}_{pq}}{\hat{b}_{pq}} \zeta_m^{pi} \zeta_n^{qj}. \tag{4.29}$$

Of course, this presumes that none of the Fourier entries of $\hat{B}$ are zero, almost always the case in practice.

So suppose that you are receiving many images over a channel with time-invariant distortions. It may be possible to discover the blurring operator B of this channel and thereafter deblur future images as in (4.29).

On the other hand, there may be reasons for intentionally blurring images, say for encryption or for reducing the bandwidth needed to transmit the image. As long as the blurring operator is known, the images may be reconstituted to their original form.

Figure 4.6. A photo is a matrix of pixels.

Example. Let us experiment with encrypting an image. A digital photo of my dog Hootie, captured with a digital camera in **jpeg** format, is stored in the working directory as the file **hootie.jpg**. The image has been cropped to 393×509 pixels. This image is imported into MATLAB with the following script. The result appears in Figure 4.6.

```
% Routine 4.2  Importing photoimages
X = imread('hootie.jpg');      % import as a matrix
X = double(X);                 % convert to real data
Y = X(:,:,1)+X(:,:,2);         % average red-green-blue to
Y = (Y(:,:)+X(:,:,3))/3;       % gray scale.
X = Y/3;                       % darken the image
X = round(X);                  % round to nearest integer
 [m,n] = size(X);              % size X for later processing
image(X)                       % show the image
axis off                       % delete the axes
colormap gray                  % convert to gray scale
```

Let us encrypt this image with the blurring matrix B = rand(m,n) of random entries known only to us and our intended correspondent.

```
% Routine 4.3  Encryption
B = rand(m,n);              % choose a blurring matrix
Xhat = fft2(X);            % transform X
Bhat = fft2(B);            % transform B
Yhat = Bhat.* Xhat;        % multiply entrywise
Y = ifft2(Yhat);           % transform back
Y = real(Y);               % delete small imag.  parts
Y = round(Y);              % round to nearest integer
                           %
To view the scrambled image,  rescale as follows:

Z = Y/(m*n);               % normalize
a = min(min(Z));           % min.  pixel intensity
Z = Z - a;                 % shift intensities down
b = max(max(Z));           % max.  pixel intensity
Z = 64*Z/b;                % rescale
image(Z)                   % display the encrypted image
colormap gray;             % go to gray scale
axis off                   % delete axes
```

The FFTs above will require some time to run since the matrix dimensions m, n are not, in general, powers of 2. The encrypted image is displayed as Figure 4.7.

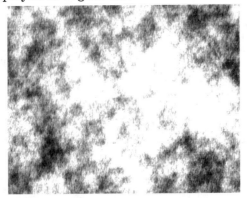

Figure 4.7. Encrypted image.

The encrypted image $Y = B * X$ depicted in Figure 4.7 is now transmitted. The receiver decodes the image back into its original by transforming to the frequency domain, dividing entrywise by $\hat{B}$, then

transforming back to the time domain by Routine 4.4. The restored
image is indistinguishable from the original.

```
% Routine 4.4  Undo encryption
Yhat = fft2(Y);              % transform blurred image
Bhat = fft2(B);              % transform blurring matrix
Xhat = Yhat./Bhat;          % divide entrywise
X1 = ifft2(Xhat);           % inverse DFT
X1 = real(X1);              % delete small imag.  parts
image(X1)                    % display reconstituted image
axis off
colormap gray
```

ALERT. In the two-dimensional circular convolution (4.25) and the
two-dimensional DFT (4.27), I have chosen to sum from 1 to m, n to
match standard matrix indexing, rather than the more conventional
sum from 0 to $m - 1, n - 1$. The results are identical when rolled
left. MATLAB uses the latter convention.

See *An Introduction to Wavelets Through Linear Algebra* by my
colleague Michael Frazier on related image processing matters, e.g.,
on the compression of the FBI fingerprint files.

Exercises

4.1 Verify Gauss's relation (4.5).

Hint: Set z equal to the left-hand side of (4.5), then compute
$\xi^a z$ in two different ways.

4.2 Prove (4.6), i.e., $F^{-1} = (1/n)\bar{F}$.

4.3 Compute the circular convolution $z = x * y$ when
$x = (1, 1, -1, -1, 2)$ and $y = (3, 1, 4, 5, 1)$.

4.4 Let $x = (x_0, x_1, \dots, x_{n-1})$ with DFT $\hat{x}_k = r_k e^{i\phi_k}$. Show that x
is real data if and only if $r_{n-k} = r_k$ and $\phi_{n-k} \equiv -\phi_k \bmod 2\pi$.

4.5*To avoid phase distortion (see Exercise 3.25), a real FIR filter

$$H(z) = h_0 + h_1 z^{-1} + \cdots + h_N z^{-N}$$

must have *constant group delay*, i.e.,

$$H(e^{i\theta}) = r(\theta)e^{i\phi(\theta)},$$

where phase lead ϕ is piecewise linear between zeros of $H(e^{i\theta})$: $\phi(\theta) = m\theta + b$. Show that in this case, for all small positive θ, $m = -N/2$ and $b = 0$ or $\pm\pi/2$, i.e.,

$$H(e^{i\theta}) = r(\theta)e^{ib - iN\theta/2}.$$

For example, the lowpass filter of Figure 4.3 is without phase distortion only if it is a filter of order $N = 1$.

Hint: Compute $H(e^{i\theta})/H(e^{-i\theta})$ in two different ways.

4.6 (**Project**) Simulate the startup time-domain response of the filter (4.13) to various sinusoidals. Redesign for better transient response.

4.7 (**Project**) "Reverse engineer" the sound of an oboe (or your choice of instrument). Experimentally add overtones to the fundamental tone and listen to the result using MATLAB's wavwrite. Adjust coefficients until you obtain a recognizable facsimile of the instrument. Or better yet, obtain a wavefile recording of the instrument with a sound card, import into MATLAB with wavread, and perform spectral analysis.

4.8 Show that the left-hand Riemann sum approximation of the kth Fourier coefficient

$$c_k = \frac{1}{2\pi} \int_0^{2\pi} x(\theta) e^{-k\theta} \, d\theta$$

for n equal subdivisions is the normalized DFT

$$\frac{\hat{x}_k}{n} = \frac{x_0 + x_1\xi^{-k} + x_2\xi^{-2k} + \cdots + x_{n-1}\xi^{-k(n-1)}}{n},$$

where $x_j = x(\theta_j)$, $\theta_j = j\Delta\theta$, $\Delta\theta = 2\pi/n$, and $\xi = e^{i\Delta\theta}$.

4.9 Revise Routine 4.1 to handle a sampling rate of $f_S = 20$ kHz and frame length $n = 8192$. Display the spectrum between 400 and 500 Hz.

4.10 Verify the inverse transform (4.27b).

4.11 Verify the transforms (4.28) after proposing a natural general-ization of circular convolution (4.9) to two dimensions.

4.12 Write a function cconv(x,y) that will perform one-dimensional circular convolution. Use the function that obtains the positive residue of x modulo m :

```
function y = mod(x,m);
y = x - ceil(x/m)*m + m;
```

4.13 Write a function cconv2(X,Y) that will perform two-dimensional circular convolution. Compare its speed against the transform method of Routine 4.3.

4.14 (**Project**) Ask your instructor to encode a plain text message by circular-convolving each $n = 5$-bit letter with a fixed integral blurring 5-vector b. Break this code!

Suggestion: So that you will have a fighting chance, agree to encode each letter in alphabetical order base 2, i.e., $A = (0,0,0,0,1), B = (0,0,0,1,0), C = (0,0,0,1,1), \dots$, with word space $(1,1,1,1,1)$.

4.15 Import a digital image into MATLAB and convert to gray scale via Routine 4.2. Experiment with pixel-by-pixel contrast en-hancement as described in §4.5.

4.16 Import two images A and B into MATLAB via Routine 4.2. Experiment with morphing the image A into the image B by making a movie of the sequence of still images $X = (1-t)A+tB$ as t increases from 0 to 1.

4.17 Guess the typical sampling rate of a modern digital oscillo-scope, then look up the specifications of the Fluke PM30XX series of instruments.

4.18 (**Project**) A spectrum/network analyzer is performing a real-time 1024-point FFT of a 100-MHz signal taken with 12-bit resolution. Estimate the time-to-multiply of the underlying processor. Defend your estimates. Compare your estimates with the specifications of Hewlett-Packard instruments.

4.19 Write a routine to perform the DFT (4.3b) directly without recourse to canned routines. Race your routine against MAT-LAB's fft.

Chapter 5
Linear Programming

A simple optimization in many variables is no longer simple.

Industrial problems are often optimization problems, many of which are linear problems. All linear optimization problems can be transformed into one standard problem form. As is traditional, the diet problem is used to illustrate the genre. You will be guided through an overview of Dantzig's famous simplex method for solving linear programming problems.

§5.1 Optimization

Many if not most problems of commerce, government, and warfare are *optimization problems*:

Maximize (or minimize) the *objective* (or *cost*) *function*

$$J = J(x_1, x_2, \ldots, x_n) \tag{5.1}$$

subject to the *constraints*

$$
\begin{aligned}
E_1(x_1, x_2, \ldots, x_n) &= 0, \\
E_2(x_1, x_2, \ldots, x_n) &= 0, \\
&\vdots
\end{aligned}
\tag{5.2a}
$$

$$
\begin{aligned}
E_r(x_1, x_2, \ldots, x_n) &= 0, \\
I_1(x_1, x_2, \ldots, x_n) &\leq 0, \\
I_2(x_1, x_2, \ldots, x_n) &\leq 0, \\
&\vdots
\end{aligned}
\tag{5.2b}
$$

$$I_s(x_1, x_2, \ldots, x_n) \leq 0.$$

Linear programming problems are optimization problems where both the objective function and all the constraints are linear. Thus we search for points x of $V = \mathbf{R}^n$ that maximize (or minimize) the linear functional

$$J(x) = c_1 x_1 + c_2 x_2 + \cdots + c_n x_n \qquad (5.3)$$

that are *feasible*, i.e., points x that satisfy all the given constraints

$$e_{11} x_1 + e_{12} x_2 + \cdots + e_{1n} x_n = a_1,$$
$$e_{21} x_1 + e_{22} x_2 + \cdots + e_{2n} x_n = a_2,$$
$$\cdot$$
$$\cdot$$
$$\cdot$$
$$e_{r1} x_1 + e_{r2} x_2 + \cdots + e_{rn} x_n = a_r,$$
$$\qquad (5.4)$$
$$i_{11} x_1 + i_{12} x_2 + \cdots + i_{1n} x_n \le b_1,$$
$$i_{21} x_1 + i_{22} x_2 + \cdots + i_{2n} x_n \le b_2,$$
$$\cdot$$
$$\cdot$$
$$\cdot$$
$$i_{s1} x_1 + i_{r2} x_2 + \cdots + i_{sn} x_n \le b_s.$$

We assume the linear programming problem is *well posed,* i.e., that the set K of feasible points is nonempty and bounded. Because the constraints are equalities or nonstrict inequalities, K is closed and hence compact. Thus because the objective function J is continuous, we are guaranteed solutions to the problem. Moreover, the set K of feasible points is *convex;* i.e., given any two points in K, the line segment joining them also lies in K. Because the values of a linear function along a line segment are extremal at the endpoints, the extreme values of a nonconstant objective J on K cannot occur at interior points and must occur on the boundary of K. If, say, the minimum value occurs at two boundary points of K, it must occur at each point of the line segment connecting them. In any case, the extreme values of J on K must occur at *extreme points* of K, i.e., points e of K that are not interior points of any line segment lying within K (Exercise 5.1). In summary,

The solution to a well-posed linear programming problem occurs at an extreme point of the compact convex set of feasible points.

A point becomes feasible exactly when it lies on all the hyper-
planes determined by the constraint equalities $e_k^T x = a_k$ and within
the closed half spaces determined by all the constraint inequalities
$i_k^T x \le b_k$. The extreme points of such a convex polyhedral set are
commonly called *vertices*. So to solve a linear programming problem, "all we have to do" is check the values at each vertex of the
feasible set. But think on such a task. With n variables and $m < n$
constraints, the number of vertices could be as large as the binomial
coefficient $n!/m!(n-m)!$ (Exercise 5.3). Some more efficient search
algorithm is needed (see §5.3).

Since the feasible points of a well-posed problem form a bounded
set, we may as well make a linear change of variables so that all
feasible solutions have nonnegative coordinates. Moreover, we may
as well adjoin s additional nonnegative *slack* variables

$$x_{n+1}, x_{n+2}, \ldots, x_{n+s}$$

so that each inequality

$$i_{k1}x_1 + i_{k2}x_2 + \cdots + i_{kn}x_n \le b_k$$

becomes the equality

$$i_{k1}x_1 + i_{k2}x_2 + \cdots + i_{kn}x_n + x_{n+k} = b_k.$$

Finally, if we are to maximize the objective J, we may choose instead
to minimize $-J$. By these tricks we can bring any linear programming
problem to **standard form:**

Minimize in $\mathbf{R}^n$ the cost

$$J = c^T x \tag{5.5a}$$

subject to

$$Ax = b, \quad \text{where } x \ge 0. \tag{5.5b}$$

§5.2 The Diet Problem

Imagine that you are a nutritionist charged with planning healthy
but economical meals for the inmates of a state prison. You must
keep cost low, yet satisfy inmate nutritional requirements. You can
purchase n foods X_j at price p_j per ounce of serving. For instance,

X_1 might be pasta at $p_1 = 2$ cents per ounce, X_2 green peas at $p_2 = 5$ cents per ounce, etc. Each food X_j has m nutritional elements a_{ij}, such as calories, fiber, vitamin C, etc. per ounce of serving. Finally, each inmate has a daily minimum requirement c_i of each of the m dietary elements (Table 5.1).

Table 5.1 ([Dorfman et al.])

Nutritional	Food				Minimum daily
element	X_1	X_2	...	X_n	requirement
1	a_{11}	a_{12}	...	a_{1n}	c_1
2	a_{21}	a_{22}	...	a_{2n}	c_2
.					
.	...	...	...	...	...
.					
m	a_{m1}	a_{m2}	...	a_{mn}	c_m

Your task is to solve for the meal portions x_j (in ounces) with minimum cost

$$J = p_1 x_1 + p_2 x_2 + \cdots + p_n x_n, \qquad (5.6)$$

subject to the dietary constraints

$$
\begin{aligned}
a_{11} x_1 + a_{12} x_2 + \cdots + a_{1n} x_n &\geq c_1, \\
a_{21} x_1 + a_{22} x_2 + \cdots + a_{2n} x_n &\geq c_2,
\end{aligned}
$$

$$\vdots \qquad\qquad (5.7a)$$

$$a_{m1} x_1 + a_{m2} x_2 + \cdots + a_{mn} x_n \geq c_m.$$

This is not a well-posed problem as yet. Adjoin the additional real-world constraints that the portions cannot be negative, i.e.,

$$x_j \geq 0. \qquad (5.7b)$$

Finally, adjoin the cost of some very expensive meal as a constraint:

$$p_1 x_1 + p_2 x_2 + \cdots + p_n x_n \leq P, \qquad (5.7c)$$

whose half plane contains at least one point satisfying all the other constraints (5.7a,b). Since each price $p_j > 0$, the problem is now well posed since the set K of feasible points satisfying (5.7) is compact and nonempty. A case where $n = 2$ and $m = 4$ is shown in Figure 5.1.

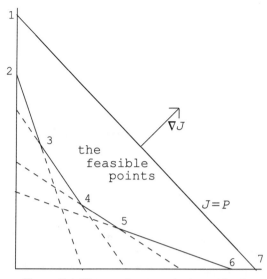

Figure 5.1. The most economical meal is at vertex 4, since it is furthest upstream of ∇J.

The negative gradient of cost $-\nabla J = -(p_1, p_2, \ldots, p_n)$ points in the direction of maximal decrease of cost J (in the general direction of the origin). So the least expensive healthy meal is found at the furthest downstream vertex of the current $-\nabla J$. In Figure 5.1 the solution is vertex 4.

§5.3 The Simplex Algorithm

It is estimated that hundreds of millions of dollars in human and computer time is invested yearly in the formulation and solution of linear programming problems [Wright]. A typical problem might have $n = 200$ variables with $m = 40$ constraints, yielding possibly $200!/40!160! \approx 2 \times 10^{42}$ vertices to check for a solution (Exercise 5.3). A great deal of thought has gone into the design of algorithms that most efficiently select and check likely vertices. The best known algorithm against which all others are judged is the *simplex method*, a method developed in 1947 by G. B. Dantzig. This method finds an extreme point, then selects another extreme point with an optimally improved objective value.

Step 1. Recast the problem into standard form (see §5.1):

Minimize in $\mathbf{R}^n$ the cost

$$J = c^T x \tag{5.8a}$$

subject to the m constraints

$$Ax = b, \tag{5.8b}$$

where

$$x \geq 0. \tag{5.8c}$$

We assume that $b \neq 0$.

Step 2. Find one feasible point.

One inefficient but often successful way to accomplish this is to solve for the slack variables with all other variables set to zero. See [Pierre] for various methods of attack.

Another sometimes successful approach is by row reducing $Ax = b$. Augment the $m \times n$ matrix A with the $m \times 1$ column b. Now apply in the usual way elementary row operations — you may switch rows, multiply rows by nonzero constants, or add a multiple of a copy of one row to another — to bring $(A|b)$ to row-reduced echelon form. Then resubscript the variables x_j to bring $(A|b)$ to the form

$$\begin{pmatrix} 1 & 0 & . & 0 & * & . & * & b'_1 \\ 0 & 1 & . & 0 & * & . & * & b'_2 \\ . & . & . & . & . & . & . & . \\ 0 & 0 & . & 1 & * & . & * & b'_r \\ 0 & 0 & . & . & . & . & . & 0 \\ . & . & . & . & . & . & . & . \\ 0 & 0 & . & . & . & . & . & 0 \end{pmatrix} . \tag{5.9}$$

Recall that elementary row operations do not disturb solutions to $Ax = b$. The set K of feasible points is unchanged — we have merely simplified the form of the constraints. If the system of equations is inconsistent (with some $b_{r+k} \neq 0$), the problem is ill posed — there are no solutions at all to $Ax = b$, not to speak of feasible solutions $x \geq 0$. If all the nonzero b'_j are positive, we have our feasible solution, viz., $x_j = b'_j$.

Even if this attempt fails, we have revealed the redundant constraints — they have become zero rows and are no constraint at all. We may as well strike out the zero rows and assume that A has rank $r = m$.

Step 3. Find an extreme (*basic*) feasible point.

From step 2 we have in hand a feasible point $x = (x_1, x_2, \ldots, x_n)^T$, a solution to (5.8b) with all $x_j \geq 0$. We will use this known feasible point to obtain a vertex (extreme point) of the set K of feasible points. The means to this end is the θ trick:

Since x satisfies (5.8b), resubscript the x coordinates so that

$$x_1 a_1 + x_2 a_2 + \cdots + x_k a_k = b, \tag{5.10}$$

where a_j is the jth column of A, where $x_j > 0$ for $j = 1, 2, \ldots, k$ and $x_j = 0$ for $j = k + 1, \ldots, n$. We may as well assume that the first r columns are independent and span the remaining $k - r$ columns. Write the column a_{r+1} in terms of the first r:

$$a_{r+1} = y_1 a_1 + y_2 a_2 + \cdots + y_r a_r \tag{5.11}$$

and substitute in (5.10) to obtain the relation

$$(x_1 + x_{r+1} y_1) a_1 + (x_2 + x_{r+1} y_2) a_2 + \cdots + (x_r + x_{r+1} y_r) a_r$$
$$+ x_{r+2} a_{r+2} + \cdots + x_k a_k = b \tag{5.12}$$

with at least one fewer nonzero x coordinate. Of course, the resulting point x' may not be feasible since several of its coordinates $x_j + x_{r+1} y_j$ may be negative.

But perform a homotopy, the θ *trick:* Form the linear combination

$$x'' = (1 - \theta)x + \theta x' \tag{5.13}$$

for $0 \leq \theta \leq 1$. Note that $Ax'' = b$. As θ slowly rises from zero, the resulting point x'' begins with k positive coordinates, but moves toward a point with less than k nonzero coordinates. So somewhere, for some first value of θ between 0 and 1, x'' has nonnegative coordinates but fewer than k positive coordinates. Thus we have in hand a feasible point with fewer than k nonzero coordinates.

Using the θ trick repeatedly, obtain a feasible point of independent columns. But these are vertices!

LEMMA. Feasible points x where

$$x_1 a_1 + x_2 a_2 + \cdots + x_r a_r = b \tag{5.14}$$

with $x_j > 0$ and $a_1, \ldots, a_r$ independent, are extreme feasible points, and conversely.

PROOF. Suppose that x is an extreme feasible point as in (5.14), yet the respective columns are dependent, i.e.,

$$y_1 a_1 + y_2 a_2 + \cdots + y_r a_r = 0. \tag{5.15}$$

Set $y = (y_1, y_2, \ldots, y_r, 0, 0, \ldots, 0)$. Then

$$x = \frac{x + \epsilon y}{2} + \frac{x - \epsilon y}{2},$$

where both terms on the right have nonnegative coordinates for all small ϵ and are thus feasible since $A(x \pm \epsilon y) = b$; i.e., x is not extreme.

Conversely, suppose that the r columns in (5.14) are independent, yet

$$x = \frac{x' + x''}{2}$$

for two feasible points x' and x''. Since each of the n coordinates of both x' and x'' are nonnegative and average to the corresponding coordinate of x, each coordinate of x' and x'' with subscript exceeding r must be zero, giving a nonunique representation

$$x_1' a_1 + \cdots + x_r' a_r = b = x_1'' a_1 + \cdots + x_r'' a_r$$

unless $x = x' = x''$. Thus x is an extreme point.

Step 4. Move to a basic (extreme) feasible point of lower cost.

Given a basic feasible point x of cost $J(x) = c^T x$, we go in the direction of *steepest descent* toward another basic feasible point x' of lower cost $J(x')$. We again employ a θ trick.

Suppose the basic feasible point x is as in (5.14) with $a_1, \ldots, a_r$ independent. We must take on an additional assumption.

NONDEGENERACY ASSUMPTION: $r = m$. Stronger yet, assume that b does not lie in the span of any $m - 1$ columns of A.

This assumption can be assured by slightly perturbing the constraints (see [Pierre]).

Write each later column

$$a_j = y_{1j} a_1 + y_{2j} a_2 + \cdots + y_{mj} a_m \tag{5.16}$$

in terms of the first m for $j = m+1, m+2, \ldots, n$. Then for each such j and each $\theta \geq 0$,

$$(x_1 - \theta y_{1j})a_1 + \cdots + (x_m - \theta y_{mj})a_m + \theta a_j = b, \qquad (5.17a)$$

with cost

$$J(x(\theta)) = c_1(x_1 - \theta y_{1j}) + \cdots + c_m(x_m - \theta y_{mj}) + c_j\theta$$

$$= J(x) - \theta(c_1 y_{1j} + \cdots + c_m y_{mj} - c_j). \qquad (5.17b)$$

So it is clear from (5.17b) in what direction away from the vertex x we must steer — in the direction where

$$d_j = c_1 y_{1j} + \cdots + c_m y_{mj} - c_j \qquad (5.18)$$

is positive and maximal among the choices $j = m+1, m+2, \ldots, n$. If the term (5.18) is never positive, **exit** the simplex method — we have our solution x (Exercise 5.10). Otherwise, once the j with maximal positive d_j is selected, slowly increase θ, keeping all coordinates in (5.17a) nonnegative until the first vanishing of a coordinate of (5.17a) occurs, i.e., when

$$\theta = \min_{i=1,\ldots,m} \frac{x_i}{y_{ij}}, \quad y_{ij} \neq 0, \qquad (5.19)$$

yielding a new basic feasible point of lower cost. [If vanishing were not to occur, then from (5.17b), the cost J could be decreased without bound; the problem is not well posed.] Thus by steepest descent, we have found a vertex of lower cost. Loop back to step 4.

The literature on this putative simple subject is vast, as befits its economic performance. See the book *Linear Programming and Economic Analysis* by the distinguished economists Dorfman, Samuelson, and Solow for applications to commerce. For mathematical details and application to control, warfare, and communication, see *Optimization Theory with Applications* by Pierre. Both books exist in low-cost Dover editions. But the clearest presentation may be the book *Methods of Mathematical Economics* by Franklin.

Exercises

5.1* The extreme (maximum and minimum) values of a well-posed linear programming problem (5.3)–(5.4) must occur at an extreme point of the set of feasible points K.

Outline: The subset K_0 of K where the extreme value is realized is itself convex and compact, and its extreme points must be extreme for K. Assume $k_0 = 0$ is a boundary point of K_0. Find a continuous linear functional η so that $\eta(K_0) \leq 0$ with kernel N. The functional η has maximal value 0 on $N \cap K_0$, which, by induction on dimension, occurs at an extreme point.

5.2* Show that the maximum of a *convex* function subject to well-posed constraints (5.4) is realized at an extreme point of the set of all feasible points K.

Definition: A real-valued function $\xi(x)$ on a convex set K is said to be *convex* if

$$\xi(ta + (1-t)b) \leq t\xi(a) + (1-t)\xi(b)$$

for all points $a, b \in K$ and $0 \leq t \leq 1$.

5.3 Show that the number of vertices of a feasible set $K \subset \mathbf{R}^n$ with m constraints can be as large as the binomial coefficient $n!/m!(n-m)!$.

Hint: See (5.9).

5.4 Verify that indeed $200!/40!160! \approx 2 \times 10^{42}$.

5.5 (**Project**) Estimate the time required with present technology to solve a linear programming problem of the size in Exercise 5.4 by brute force — by finding and checking each vertex.

5.6 Minimize $J = x + 3y$ subject to $x \geq 2$, $y \geq 1$, $x + 2y \leq 8$, $x + y \leq 6$.

5.7 Minimize $J = x + 2y + 3z$ in the first octant subject to $y - x \geq 0$, $x + z \geq 1$.

5.8 Solve the following diet problem:

	X_1	X_2	Min. req.
E_1	10	4	40
E_2	7	7	49
E_3	5	10	50
Price	5	7	

.

5.9 Show that for convex domains K, local extrema of linear functions are global extrema.

5.10*Prove that if no d_j of (5.18) is positive, the linear programming problem is solved.

Outline: Perturb the vertex x to a nearby feasible point

$$x' = (x_1 + \theta_1, \ldots, x_m + \theta_m, \theta_{m+1}, \ldots, \theta_n),$$

where the $\theta_j \geq 0$ for $j = m+1, \ldots, n$. Using (5.16) and $d_j \leq 0$ from (5.18), deduce that $J(x) \leq J(x')$. Apply Exercise 5.9.

5.11 (**Project**) Report on what is publicly known about the proprietary *interior point method* of Karmarkar for solving linear programming problems.

5.12 To each *primal* linear programming problem: *minimize $J(x) = c^T x$ subject to the constraints $Ax \geq b$, $x \geq 0$*, there can be associated the *dual* problem: *maximize $J^*(y) = b^T y$ subject to the constraints $A^T y \leq c$, $y \geq 0$.*

Prove that if x and y are feasible for their respective problems, i.e., satisfy their respective constraints, then

$$J^*(y) \leq J(x).$$

5.13 Prove that if x_0 and y_0 are feasible for the primal and dual problems, respectively, with $J^*(y_0) = J(x_0)$, then both x_0 and y_0 solve their respective linear programming problem.

5.14**Show conversely that if x_0 and y_0 solve the primal and dual problem, respectively, then $J^*(y_0) = J(x_0)$. Assume both problems are well posed, so that the set of feasible points for both are compact (see [Franklin]).

5.15**Assuming that both problems are well posed, show that if the primal problem is solvable, so is its dual.

5.16 Write down the problem dual to the diet problem of Exercise 5.8. What are the units of the dual variables y?

5.17 (Ricardo's theory of comparative advantage) Each of m countries can produce x_{ij} units of the jth commodity X_j with a certain efficiency a_{ij} (in dollars per unit), i.e.,

$$a_{i1}x_{i1} + a_{i2}x_{i2} + \cdots + a_{in}x_{in} \leq b_i,$$

where b_i denotes the total productive resources of the ith country. Free unrestricted international trade yields a price p_j for each commodity X_j. Thus the *national product*, the wealth of country i, is

$$W_i = p_1 x_{i1} + p_2 x_{i2} + \cdots + p_n x_{in}.$$

Argue that the barter price p_j (in dollars per unit) of the commodity X_j satisfies $\min_i a_{ij} \leq p_j \leq \max_i a_{ij}$. Prove that each country should specialize in the one commodity that it produces most efficiently.

5.18 Each of three products X_1, X_2, X_3 can be sold for \$1 per unit. The products are produced with one of two machines, M_1 or M_2, with different output rates per 1000 units:

	X_1	X_2	X_3
M_1	4	2	3
M_2	3	6	5

e.g., the second machine can produce 1000 units of product 2 in 6 days but requires only 5 days for product 3. What mix of product yields 1000 units in the least time? Assume each product will be assigned to only one machine.

5.19 Solve Exercise 5.18 assuming that each product can be apportioned out to both machines.

5.20 Under the assumptions of Exercise 5.19 except that that product X_i sells for i dollars per unit, find the strategy that maximizes income per unit time.

Chapter 6
Regression

Data is best fit by minimizing least square error.

Which curve best fits data obtained from an experiment? We begin by fitting discrete data and are led to *regression* — approximation in the least square sense. The various norms on $\mathbf{R}^n$ are introduced only to be shown equivalent. Least square error arises from the concept of *inner product*, giving rise to Hilbert space. We introduce several application-important Hilbert spaces. The Hilbert space viewpoint unifies the fitting of discrete and continuous data into one context. We end with an application to electric power quality and a brief introduction to Fourier series.

§6.1 Best Fit to Discrete Data

Suppose we have taken n experimental data points

x	x_1	x_2	$\ldots$	x_n
y	y_1	y_2	$\ldots$	y_n

and have good reason to believe that this data, were it ideal, would follow some analytic rule

$$y = \psi(x, a), \tag{6.1}$$

where a is a vector of parameters to be adjusted until the fit is best, in some sense.

Example 1. What line $y = mx + b$ best fits the data

x	1.0	1.7	2.3	3.1
y	0.6	1.4	3.2	5.2

?

Here the parameters are the slope m and intercept b.

There are three reasonable interpretations of *best fit.*

INTERPRETATION A. Choose the parameters a to minimize the largest y-value error

$$e = \max_{j=1,\dots,n} |y_j - \psi(x_j, a)|. \tag{6.2}$$

INTERPRETATION B. Choose the parameters a to minimize the sum of the absolute errors

$$e = \sum_{j=1}^{n} |y_j - \psi(x_j, a)|. \tag{6.3}$$

INTERPRETATION C. Choose the parameters a to minimize the sum of the square errors

$$e = \sum_{j=1}^{n} [y_j - \psi(x_j, a)]^2. \tag{6.4}$$

In monetary matters and heat transfer you might prefer Interpretation B. In electronics you would choose C since energy per unit time is proportional to the square of voltage. Interpretations A and B are mathematically intractable (Exercise 6.3) and are relegated to very specialized problems. *For the remainder of this book, best fit will be in the mean square sense C.*

Example 2. What line $y = mx + b$ best fits the data

x	x_1	x_2	$\dots$	x_n
y	y_1	y_2	$\dots$	y_n

?

The mean square error is

$$e(m, b) = \sum_{j=1}^{n} (mx_j + b - y_j)^2. \tag{6.5}$$

We minimize this error by solving for the minimum of this quadratic function $e(m, b)$ of the two real variables m and b. Local minimums will occur only where the partial derivatives of e with respect to both m and b vanish, i.e., only where

$$\frac{\partial e}{\partial m} = 2 \sum_{j=1}^{n} (mx_j + b - y_j)x_j = 0 \tag{6.6a}$$

and

$$\frac{\partial e}{\partial b} = 2\sum_{j=1}^{n}(mx_j + b - y_j) = 0, \tag{6.6b}$$

which become the system of two linear equations in the unknowns m and b :

$$m\Big(\sum_{j=1}^{n} x_j^2\Big) + b\Big(\sum_{j=1}^{n} x_j\Big) = \sum_{j=1}^{n} x_j y_j \tag{6.7a}$$

and

$$m\Big(\sum_{j=1}^{n} x_j\Big) + bn = \sum_{j=1}^{n} y_j. \tag{6.7b}$$

The solution of these simultaneous equations for m and b will yield only one candidate for best-fitting line, which is indeed the minimal error since the *Hessian*

$$H(m, b) = \left(\begin{array}{cc} e_{mm} & e_{mb} \\ e_{bm} & e_{bb} \end{array}\right) \tag{6.8}$$

is positive definite (Exercise 6.1), where e_{mm} denotes the second partial of e with respect to m twice, and so on. Thus by the second derivative test, the m and b obtained from (6.7) yield the best fit.

THEOREM A. As long as ψ is differentiable in the parameters a, the best fit of $y = \psi(x, a)$ to data will occur only where

$$\frac{\partial e}{\partial a_i} = 2\sum_{j=1}^{n}[\psi(x_j, a) - y_j]\frac{\partial \psi(x_j, a)}{\partial a_i} = 0 \tag{6.9}$$

for each of the m parameters a_i that make up the vector a.

Example 3. Suppose we have reason to believe that the data

x	x_1	x_2	$\cdots$	x_n
y	y_1	y_2	$\cdots$	y_n

is ideally exponential,

$$y = ce^{kx}. \tag{6.10}$$

A frontal attack applying (6.9) would not be wise (Exercise 6.4). Instead, transform the data into logarithmic data

x	x_1	x_2	$\cdots$	x_n
$\ln y$	$\ln y_1$	$\ln y_2$	$\cdots$	$\ln y_n$

and apply linear regression (6.7).

Example 4. Which linear combination of the trial functions $\phi_j(x)$

$$\psi(x, a) = a_1\phi_1(x) + a_2\phi_2(x) + \cdots + a_m\phi_m(x) \tag{6.11}$$

best fits the data

x	x_1	x_2	$\cdots$	x_n
y	y_1	y_2	$\cdots$	y_n

?

Applying (6.9), we see that

$$\sum_{j=1}^{n}[y_j - \sum_{k=1}^{m} a_k\phi_k(x_j)]\phi_i(x_j) = 0, \tag{6.12}$$

i.e., in matrix form (Exercise 6.7),

$$\Phi\Phi^T a = \Phi y \tag{6.13}$$

where Φ is the $m \times n$ matrix $\Phi = (\phi_i(x_j))$ and where a and y are the columns $a = (a_1, a_2, \ldots, a_m)^T$ and $y = (y_1, y_2, \ldots, y_n)^T$. The system (6.13) will have a *unique* solution for the optimal parameters a when the symmetric positive semidefinite $\Phi\Phi^T$ is positive definite and hence invertible. Invertible or not, the best fit is always achieved, and at the solutions of the system (6.13) — see Gram's theorem on regression in §6.4.

Example 5. (Polynomial regression) A particular instance of Example 4 is when the trial functions are powers of x, i.e., when we attempt to fit data with polynomials of degree m:

$$\psi(x, a) = a_0 + a_1 x + a_2 x^2 + \cdots + a_m x^m. \tag{6.14}$$

In this case the matrix $\Phi\Phi^T$ of (6.13) becomes the product of two van der Monde–like matrices

$$\Phi\Phi^T = (x_j^i)(x_i^j) = (\sum_{k=1}^{n} x_k^{i+j}), \tag{6.15}$$

giving a matrix that is invertible for $n > m$ (Exercise 6.8).

§6.2 Norms on $\mathbf{R}^n$

A *norm* on a real vector space V is a mapping $v \mapsto \|v\|$ from V to the nonnegative reals with the three properties

$$\|v\| \geq 0 \text{ with equality exactly when } v = 0, \qquad (6.16a)$$

$$\|cv\| = |c| \cdot \|v\|, \qquad (6.16b)$$

$$\|u + v\| \leq \|u\| + \|v\|, \qquad (6.16c)$$

for all (real) scalars c and vectors u and v in V.

For example, on the n-tuples $x = (x_1, x_2, \ldots, x_n)$ of $\mathbf{R}^n$, there are three commonly applied norms (Exercise 6.12):

$$\|x\|_\infty = \max_{i=1,\ldots,n} |x_i|, \qquad (6.17)$$

$$\|x\|_1 = \sum_{i=1}^{n} |x_i|, \qquad (6.18)$$

$$\|x\|_2 = \left(\sum_{i=1}^{n} x_i^2 \right)^{1/2}. \qquad (6.19)$$

The last is, of course, the familiar distance formula for the length of the vector v. There are many other possible norms (Exercise 6.13), making the following result all the more astonishing.

THEOREM B. All norms on a real vector space V of dimension n are equivalent, i.e., all yield the identical convergent sequences. In fact, given any two norms, $\| \cdot \|$ and $\| \cdot \|'$ on V, each will *dominate* the other, i.e., there are positive constants c and c' such that

$$c\|v\| \leq \|v\|' \leq c'\|v\| \qquad (6.20)$$

for all v in V.

PROOF. Any norm $\| \cdot \|$ on $V = \mathbf{R}^n$ is dominated by $\| \cdot \|_\infty$ (Exercise 6.15).

Conversely, we must show that any given norm $\| \cdot \|$ on V dominates $\| \cdot \|_\infty$. Suppose that the coordinates of all the points of the unit sphere $\|x\| = 1$ are bounded in absolute value by, say, R. Thus

$$\|x\|_\infty \leq R\|x\|$$

for all points x on the sphere $\|x\| = 1$, hence for all points by homogeneity.

If, say, all the last coordinates x_n of points x on the unit sphere $\|x\| = 1$ form an unbounded set, there is a sequence of such x with $1/|x_n| = \|x/x_n\| = \|(z,1)\| \to 0$. But $\|(z,1)\| \geq |\ \|(z,0)\| - \|(0,1)\|\ |$ giving a bounded sequence of such $(z^{(k)}, 0)$. But by induction on dimension, the norm $\|z\|' = \|(z,0)\|$ on points of this $(n-1)$-dimensional subspace is equivalent to all other norms, giving that bounded sequences possess convergent subsequences. So we may as well assume $(z^{(k)}, 0) \to (z^{(0)}, 0)$. But this means

$$(z^{(k)}, 1) = (z^{(0)}, 1) + (z^{(k)} - z^{(0)}, 0),$$

where the first and last terms go to zero, an impossibility.

Thus *forcing mean square error to zero forces a best fit in every sense.* Unless explicit mention is made otherwise, we will always employ the norm $\|\cdot\|_2$, the common distance formula (6.19).

§6.3 Hilbert Space

The familiar distance formula norm (6.19) is not only physically appealing but is the most mathematically tractable since it arises from an inner product.

DEFINITION. An *inner product* on a real vector space V is a symmetric, positive definite bilinear form $\langle \cdot, \cdot \rangle : V \times V \to \mathbf{R}$, i.e.,

$$\langle u, v \rangle = \langle v, u \rangle, \tag{6.21a}$$

$$\langle v, v \rangle \geq 0 \text{ with equality exactly when } v = 0, \tag{6.21b}$$

$$\langle u + v, w \rangle = \langle u, w \rangle + \langle v, w \rangle, \tag{6.21c}$$

$$\langle cu, v \rangle = c \langle u, v \rangle, \tag{6.21d}$$

for all real scalars c and vectors u, v, w in V.

The norm arising from this inner product is

$$\|v\| = \sqrt{\langle v, v \rangle}. \tag{6.22}$$

To see that this is indeed a norm, we must check the three axioms of (6.16), where only the third presents any difficulty. The crucial missing fact is the single most useful inequality of mathematics.

The Cauchy–Schwarz Inequality. For any inner product,

$$\langle u, v \rangle \leq \|u\| \cdot \|v\| \tag{6.23}$$

with equality exactly when u and v are scalar multiples of one another.

PROOF. Expand out $\langle u - cv, u - cv \rangle \geq 0$ and set $c = \|u\|/\|v\|$ (Exercise 6.16).

A real vector space X with an inner product under which X is *complete*, i.e., all Cauchy convergent sequences converge, is called a **Hilbert space**, the most important mathematical object of the twentieth century.

A sequence of vectors $\phi_1, \phi_2, \phi_3, \dots$ from a Hilbert space X is an *orthogonal* sequence if the terms are mutually orthogonal, i.e.,

$$\langle \phi_i, \phi_j \rangle = 0 \text{ if } i \neq j.$$

A sequence $\phi_1, \phi_2, \phi_3, \dots$ is *orthonormal* if

$$\langle \phi_i, \phi_j \rangle = \begin{cases} 1 & \text{if } i = j \\ 0 & \text{otherwise.} \end{cases} \tag{6.24}$$

An orthonormal sequence $\phi_1, \phi_2, \phi_3, \dots$ is an *orthonormal basis* for the Hilbert space X if every element f of X can be written as a norm convergent series

$$f = \sum_n c_n \phi_n. \tag{6.25}$$

We restrict our attention to *separable* Hilbert spaces, spaces that possess a finite or countably infinite orthonormal basis. These are the spaces that arise physically in quantum mechanics, heat transfer, and distributed vibrations (see Chapter 11).

Example 6. $\mathbf{R}^n$ is a Hilbert space under the ordinary dot product

$$\langle x, y \rangle = \sum_{i=1}^{n} x_i y_i \tag{6.26a}$$

with norm

$$\|x\|_2 = \Big(\sum_{i=1}^{n} x_i^2 \Big)^{1/2}. \tag{6.26b}$$

Example 7. Take Example 6 to the limit: Consider all ∞-tuples of real numbers

$$x = (x_1, x_2, x_3, \dots) \tag{6.27a}$$

that are *square summable*, i.e.,

$$\sum_{i=1}^{\infty} x_i^2 < \infty. \tag{6.27b}$$

All such sequences form a complete normed vector space arising from the inner product

$$\langle x, y \rangle = \sum_{i=1}^{\infty} x_i y_i \tag{6.27c}$$

under the norm

$$\|x\| = \left(\sum_{i=1}^{\infty} x_i^2\right)^{1/2} \tag{6.27d}$$

(Exercise 6.19). This Hilbert space is sometimes called $\mathbf{R}^\infty$ but more often $\mathbf{l}^2$, pronounced "little ell two." The obvious orthonormal basis is the *natural basis*

$$e_i = (0, 0, 0, \dots, 1 \ (i\text{th position}), 0, 0, \dots).$$

Example 8. Let Ω be a bounded domain, an open connected subset of $\mathbf{R}^n$. The set

$$X = L^2(\Omega) \tag{6.28a}$$

of all square-integrable, real-valued functions f, i.e., functions f with

$$\int_\Omega f(x)^2 \, dx < \infty, \tag{6.28b}$$

forms a separable Hilbert space under the inner product

$$\langle f, g \rangle = \int_\Omega f(x)g(x) \, dx \tag{6.28c}$$

and L^2 norm

$$\|f\|_2 = \left(\int_\Omega f(x)^2 \, dx\right)^{1/2}. \tag{6.28d}$$

Since $X = L^2(\Omega)$ possesses a countably infinite orthonormal basis, it is structurally indistinguishable from $\mathbf{l}^2$. The verification of all this requires several deep analytic results (see [Rudin], [Halmos], or [Young]).

Thus a complete list of separable real Hilbert spaces is

$$\mathbf{R}, \mathbf{R}^2, \mathbf{R}^3, \mathbf{R}^4, \ldots, \mathbf{l}^2.$$

§6.4 Gram's Theorem on Regression

THEOREM C. (Gram) Let $f, \phi_1, \phi_2, \ldots, \phi_n$ be any $n+1$ elements of a real Hilbert space X. Then the best approximation of f using elements

$$\psi = c_1\phi_1 + c_2\phi_2 + \cdots + c_n\phi_n \tag{6.29}$$

drawn from the subspace S spanned by the ϕ_i is in fact achieved. This best fit is realized exactly when the coefficients c_i satisfy the n simultaneous linear equations

$$\sum_{j=1}^{n} c_j \langle \phi_i, \phi_j \rangle = \langle \phi_i, f \rangle \tag{6.30}$$

for $i = 1, 2, 3, \ldots, n$.

PROOF. The minimum error $\|f - \psi\|$ is achieved at some $\psi \in S$ since S is finite-dimensional. To see this, suppose that ψ_k are chosen from S so that $\|f - \psi_k\|$ decreases to the infimum of $\|f - \psi\|$ over S. Then since the sequence of ψ_k is bounded, it possesses a convergent subsequence that converges within S to an element ψ_0 that realizes the least possible error

$$e = f - \psi_0$$

in norm.

Write this best fit in terms of the trial functions

$$\psi_0 = c_1\phi_1 + c_2\phi_2 + \cdots + c_n\phi_n.$$

We now employ a standard argument from the calculus of variations [Courant]. For any element ϕ of S,

$$\langle e, e \rangle \le \langle e + \epsilon\phi, e + \epsilon\phi \rangle = \langle e, e \rangle + 2\epsilon\langle e, \phi \rangle + \epsilon^2\langle \phi, \phi \rangle. \tag{6.31}$$

But for very small ϵ, the term $2\epsilon\langle e, \phi\rangle$ in (6.31) dominates the ϵ^2 term to yield a value $\langle e + \epsilon\phi, e + \epsilon\phi\rangle$ or $\langle e - \epsilon\phi, e - \epsilon\phi\rangle$ even smaller than $\langle e, e\rangle$ unless

$$\langle e, \phi\rangle = 0 \tag{6.32}$$

for all ϕ in the subspace S. Thus *the best-fit error is orthogonal to all the elements of the approximating subspace* (see Figure 6.1). This means $\langle \phi_i, e\rangle = 0$ for each i, which is (6.30) in disguise.

Conversely, any solution of (6.30) yields an error $e = f - \psi_0$ orthogonal to S that by (6.31) is the minimal error. Moreover, ψ_0 is unique.

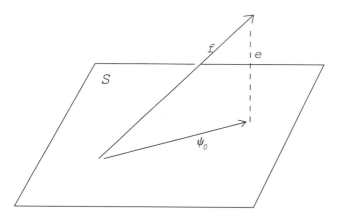

Figure 6.1. The best approximation ψ_0 to f from a subspace S is when the error e is orthogonal to S.

COROLLARY. (Bessel's theorem on regression) Let f be any element and $\phi_1, \phi_2, \phi_3, \dots$ any orthonormal sequence in the Hilbert space X. Then the least error in the approximation

$$\left\| f - \sum_n c_n \phi_n \right\| \tag{6.33}$$

will occur when each

$$c_n = \langle f, \phi_n\rangle. \tag{6.34}$$

Moreover, we have *Bessel's inequality*,

$$\langle f, f\rangle \geq \sum_n \langle f, \phi_n\rangle^2. \tag{6.35}$$

PROOF. The formula (6.34) for the optimal coefficients follows from Gram's theorem. Once choices (6.34) are made, the inequality (6.35) follows by expanding out

$$\langle f - \sum_n c_n \phi_n, f - \sum_n c_n \phi_n \rangle \geq 0. \tag{6.36}$$

Problem. Computer switching-mode power supplies, variable-speed motor controllers, and other nonlinear devices connected to the electric power mains are creating serious industrial *power quality* problems. The resulting distorted waveforms appear as sums of sinusoidals that best fit the distorted waveform in energy.

Find the energy best fit to the periodic square-wave voltage

$$f(t) = \text{sgn}(\sin t) \tag{6.37}$$

of the sinusoids

$$\psi(t) = a_0 + \sum_{n=1}^{5} a_n \cos nt + b_n \sin nt. \tag{6.38}$$

Solution. Electric energy is proportional to the time integral of the square of the voltage (or current). Thus energy best fit is best fit in the L^2 sense.

By direct calculation (Exercise 6.22) the 11 approximating trial functions

$$1, \ \cos t, \ \sin t, \ \cos 2t, \ \ldots, \sin 5t \tag{6.39}$$

are mutually orthogonal with respect to the inner product

$$\langle f, g \rangle = \int_{-\pi}^{\pi} f(t)g(t) \, dt \tag{6.40}$$

on the Hilbert space $X = L^2(-\pi, \pi)$. Thus the Gram system of equations (6.30) are diagonal, i.e., if the approximation (6.38) is the best fit, then

$$a_0 = \frac{\langle f, 1 \rangle}{\langle 1, 1 \rangle} = \frac{1}{2\pi} \int_{-\pi}^{\pi} f(t) \, dt = 0, \tag{6.41}$$

and for $n > 0$,

$$a_n = \frac{\langle f, \cos nt \rangle}{\langle \cos nt, \cos nt \rangle} = \frac{1}{\pi} \int_{-\pi}^{\pi} f(t) \cos nt \, dt = 0, \tag{6.42}$$

$$b_n = \frac{\langle f, \sin nt \rangle}{\langle \sin nt, \sin nt \rangle} = \frac{1}{\pi} \int_{-\pi}^{\pi} f(t) \sin nt \, dt = 2\frac{1 - (-1)^n}{n\pi} \quad (6.43)$$

(Exercise 6.23). Thus the best fit to this periodic waveform f is its *Fourier series* (see Chapter 11).

In summary, to best fit *continuous* data $f(x)$ (for x ranging over a domain Ω), by functions of some special form $\psi(x, a)$, search for the parameters a that yield the least error

$$\|e\|_2 = \|f(x) - \psi(x, a)\|_2$$

in the Hilbert space $X = L^2(\Omega)$.

To best fit discrete data $y_k = f(x_k)$ by $y = \psi(x, a)$, search for the parameters a that yield the least error

$$\|e\|_2 = \|(f(x_1) - \psi(x_1, a), f(x_2) - \psi(x_2, a), \ldots, f(x_n) - \psi(x_n, a))\|_2$$

in the Hilbert space $X = \mathbf{R}^n$.

Exercises

6.1 Give a commonsense argument that the line $y = mx + b$ determined by (6.7) is indeed the global minimum, hence the best fit to the data. Then supply an analytic proof using the Hessian.

6.2 By hand calculation find the line that best fits the four data points of Example 1.

6.3 Take the data of Example 1 and find the line $y = mx + b$ that best fits the data in the sense of Interpretation B of (6.3). Notice how the absolute value ruins the obvious approach. Absolute value is not differentiable.

6.4 Attempt to exponentially fit (6.10) using (6.9). How would you solve the resulting equations for c and k? Challenge *Mathematica* with a small data set.

6.5 Write a MATLAB script to do linear regression (6.7).

6.6 Find the best-fitting quadratic $y = ax^2 + bx + c$ to the data

x	-2	-1	1	2	3
y	2	1	-1	0	2

Graph the results.

6.7 Verify (6.12) and (6.13).

6.8*Show that $\Phi\Phi^T$ in (6.15) is invertible when $n > m$ since the abscissas x_k of given data are always by assumption distinct.

6.9 Write a MATLAB routine for best fitting a polynomial of degree m to data points (x_j, y_j) for $j = 1, 2, 3, \ldots, n$, where $n > m$ and where the x_j are distinct. Compare your results against MATLAB's `polyfit`.

6.10 (Project) Using census data from 1790 to today, experimentally find a best-fitting polynomial for U.S. population in millions against time t in decades.

6.11 (Project) State space systems that evolve in time are often modeled using their *proper orthogonal decomposition* (POD) modes ϕ_k:

$$X(t) = c_1(t)\phi_1 + c_2(t)\phi_2 + \cdots + c_m(t)\phi_m.$$

The POD modes ϕ_k are determined statistically as follows: At the jth sample time t_j, measure the state

$$X_j = (x_1(t_j), x_2(t_j), \ldots, x_m(t_j))^T$$

of the system, a $m \times 1$ column of readings. Take n samples with $n \gg m$. Then form the $m \times m$ matrix

$$A = \frac{1}{n}\Phi\Phi^T \approx E[X \cdot X^T],$$

where $\Phi = (x_i(t_j))$ is the $m \times n$ matrix of state columns X_j. The m eigenvectors ϕ_i belonging to the (nonnegative) eigenvalues of A (the squares of the singular values of Φ) form an orthogonal basis for state space.

Argue that these POD modes form the natural basis for the time evolving states of the system since they arise from statistical data of actual performance. Now take the other side and argue that the POD modes are irrelevant since they are not based on physical law — after all, you could adjoin (say) a commodity price as an additional $(m + 1)$th state.

Experiment with a phase space proper orthogonal decomposition of a simulated one-mass, two-spring, or a two-mass, three-spring mechanical system (see Figure 9.7). Add noise with

MATLAB's `randn`, and compute the autocorrelation modes ϕ_i. Do you recover physically meaningful eigenvalues and modes?

6.12 Prove that each of (6.17)–(6.19) is a norm on $\mathbf{R}^n$.

6.13 Find infinitely many additional novel distinct norms on $\mathbf{R}^n$.

6.14 Find two positive constants c and c' such that

$$c\|v\|_2 \le \|v\|_1 \le c'\|v\|_2$$

for all v in $\mathbf{R}^n$.

Answer: $c = 1,\ c' = \sqrt{n}$.

6.15 Using the triangle inequality, show that every norm on $\mathbf{R}^n$ is dominated by $\|\cdot\|_1$, which is in turn dominated by $\|\cdot\|_\infty$.

6.16 Complete all details in the proof of the Cauchy–Schwarz inequality (6.23).

6.17 Deduce the triangle inequality (6.16c) from the Cauchy-Schwarz inequality (6.23).

6.18 Deduce the completeness of $\mathbf{R}^n$ from the completeness of $\mathbf{R}$.

6.19* Show that $X = l^2$ of Example 7 is indeed a Hilbert space, i.e., that X is a vector space, that the form of (6.27c) is an inner product on X, and that X is complete.

6.20 (**Project**) Using Gram's theorem on regression, find in the Hilbert space $X = L^2(0,1)$ (the space of all square-integrable functions on the unit interval $[0,1]$) the best-fitting polynomials of degrees 2,3,4,5 to $f(x) = e^{-x}$. Write a MATLAB script to solve the resulting system of equations (6.30) using the command `x = A\b` that solves $Ax = b$. Graph your best approximation against the exact function f and the Taylor polynomial for f of the corresponding degree. Compute the ratio of errors of the two approximations. You will be amazed. But now try higher and higher degree polynomial approximations. Something seriously goes wrong. What has happened?

Hint: Look up the concept *condition number*.

6.21 Provide an independent proof of Bessel's Theorem on Regression of §6.4 by expanding out

$$\langle f - \sum_n c_n \phi_n, f - \sum_n c_n \phi_n \rangle \geq 0$$

and completing the square.

6.22 Show that the list of functions (6.39) are all mutually orthogonal.

6.23 Verify each equality in (6.41), (6.42), and (6.43).

6.24 Graph the best approximation ψ of the form (6.38) [with coefficients given by (6.41)–(6.43)] to the square wave f of (6.37) against f over one period. Numerically estimate the energy error $\|f - \psi\|^2$. What percentage of the total energy is this error?

6.25 (**Project**) Obtain the average temperature for each day over one year. Fit this data with a sum of sinusoids.

6.26 (**Project**) Fit some period of fluctuating commodity prices with a sum of sinusoids. Do you see any patterns?

6.27 Deduce from Gram's theorem that (6.13) has solutions.

6.28*(**Project**) Carefully redo the definitions and results of this chapter for complex data y, complex-valued approximators $\psi(a, x)$, and vector spaces with complex scalars.

6.29*Sample the signal $x(t) = f(t)$ in intervals of time $T = 2\pi/N$ to obtain the discrete data

t	t_0	t_1	$\cdots$	t_{N-1}
x	x_0	x_1	$\cdots$	x_{N-1}

Obtain the best-fitting approximation $\psi(t, a)$ of the form

$$\psi(t, a) = a_0 + a_1 e^{tTi} + a_2 e^{2tTi} + \cdots + a_{N-1} e^{(N-1)tTi}.$$

Answer: The discrete Fourier transform

$$a_k = \sum_{j=0}^{N-1} x_j \xi^{-jk}, \quad \text{where } \xi = e^{Ti}.$$

6.30 Let $\{\phi_n\}_{n=1}^{\infty}$ be any orthonormal sequence and f an arbitrary element of a Hilbert space. Prove $\langle f, \phi_n \rangle \to 0$.

Hint: Apply Bessel's inequality and the Nth term test.

6.31 Find the best fit to $f(x) = x$ on $0 < x < 1$ of the form

$$\psi(x) = \sum_{n=1}^{N} c_n \sin n\pi x.$$

6.32 Show that the $\phi_n = \sqrt{2} \sin \pi x$ form an orthonormal sequence in the Hilbert space $X = L^2(0, 1)$ under its natural inner product (6.28).

6.33 Find the best polynomial fit of degree $n = 4$ to $f(x) = |x|$ on $[-1, 1]$. Display the fit graphically. Experiment by increasing the degree n. Does the error continue to decrease?

6.34 Compute the norm of x^n in the Hilbert space $X = L^2(-1, 1)$.

6.35 Estimate the norm of e^{-x^2} in the Hilbert space $X = L^2(0, 1)$.

Answer: ≈ 0.773398.

Chapter 7
Cost-Benefit Analysis

Investment is rationally judged via discounted cash flow.

Can investment decisions in machinery, human resources, or infrastructure be put on a rational basis? You will learn the decision method of life-cycle costing via the present value of future money.

§7.1 Present Value

What is future money worth to you today? An inheritance is to be signed over to you on your thirtieth birthday. How much will a sensible banker loan you with that future sum as security?

The present worth of a future sum is what will grow to the future sum in the time allotted.

Example. An inheritance of $p = \$100,000$ is due in $N = 10$ years. Its present value p_0 is then the amount that will grow to $p = \$100,000$ when placed into an absolutely safe investment compounding continuously at, say, $r = 0.05 = 5\%$ per year for N years; i.e.,

$$p_0 e^{rN} = p_0 e^{0.5} = 100,000,$$

i.e., $p_0 = \$60,653.07$.

In summary, the present value p_0 of future money p that is paid N years in the future is

$$p_0 = p e^{-rN}. \tag{7.1}$$

The interest rate r of this fictitious absolutely safe investment in (7.1) is called the *discount rate,* a number chosen carefully after many considerations (see [EPRI, p. 2–3]). The discount rate r is often close to the projected return from U.S. Treasury notes. A high discount rate degrades the value of future income, savings, or losses.

Problem. The Michigan lottery offers two payout plans to winners: plan A, where half is paid in one lump sum; and plan B, where the full amount is paid in equal installments over 25 years. Which plan is more advantageous to the winner?

Solution. Suppose that the prize is p. Assume a discount rate of r. The present value of p in plan A is $0.5p$ and is thought of as the (lost) *opportunity cost* to plan B. The present value of the return under Plan B is

$$\frac{p}{25} + \frac{p}{25}e^{-r} + \frac{p}{25}e^{-2r} + \cdots + \frac{p}{25}e^{-24r} = \frac{p}{25}\frac{1 - e^{-25r}}{1 - e^{-r}}.$$

The present worth of these two plans agree when

$$12.5(1 - e^{-r}) = 1 - e^{-25r}, \tag{7.2}$$

which implies a discount rate of $r = 0.0673745$ (Exercise 7.1). Thus the winner would need an absolutely guaranteed return on investment of at least 6.7% before plan A becomes the better choice. The state of Michigan (and the IRS) would prefer that you chose plan A (see Exercise 7.2). Tax considerations would tilt the decision even more toward plan B.

§7.2 Life-Cycle Savings

How does one decide whether future savings justify the purchase of newer, more efficient equipment? The agreed approach is the method of *life-cycle costing*, sometimes called *life-cycle savings* or *discounted cash flow analysis*.

DISCOUNTED CASH FLOW ANALYSIS:

Step 1. For each year of the projected life of the equipment, compute the projected savings less costs for that year, the yearly *net cash flow*.

Step 2. Discount the yearly net cash flow back to its present value.

Step 3. Sum all the net cash flow present values over the lifetime and subtract the first costs to obtain the *life-cycle savings*. If the life-cycle savings are positive, the investment is worthwhile.

The *first costs* are the marginal (additional) costs for the new equipment. The *return on investment* is the discount rate r that yields zero life-cycle savings. The *payback period* is usually taken to be the time T before the accumulating present value of net savings surpasses the initial investment (marginal first costs).

Problem. A 10-ton (120,000-Btu/hr) variable-speed heat pump is known to save yearly on average 13,533 kWh over a fixed-speed machine for locations near Atlanta, Georgia. The discount rate less projected inflation is $d - i = 2\%$. The projected commercial electric rate for the region is \$0.0777 per kWh with zero rate creep over inflation [Energy Information Administration]. The projected life of the machine is 12 years. What additional first costs are justified for purchasing such an advanced machine? (From [McCleer et al.].)

Solution. The marginal saving over standard equipment in the kth year is rate $\times$ energy/yr $= 0.0777 \times 13,533 = \$1051.51$ inflated by e^{ki}. This saving must be discounted back to present value by the factor e^{-kd}. Thus since $d - i = 0.02$, the total present worth of the savings over the lifetime is

$$1051.51\left(e^{i-d} + e^{2(i-d)} + \cdots + e^{12(i-d)}\right) \approx 11,106.34$$

so you are justified in paying \$11,106 more for the advanced machine. Analysis of this sort is most often reported in tabular form as shown in Table 7.1. A spreadsheet is the tool of choice for these problems.

It is fortunate that the calculation of present worth incorporating the effects of inflation depends on the *difference* between the discount and inflation rates since this difference is historically more stable than either term forming the difference.

All this so far falls under the rubric *engineering economy*, where savings and costs are clearly quantifiable. Any engineering library will stock many texts on the subject (e.g., [Steiner]). *Cost-benefit analysis* is a term applied to grander investment decisions, where political, economic, societal, international, environmental, limited resources, and legal issues come into play. Shall we lower the water level of the upper Great Lakes by 2 feet? Should class size be reduced nationwide in the public schools? Should we colonize Mars? Should passenger rail service be expanded in the northeastern corridor? The overall approach remains the same — capture and quantify benefits

less costs, then discount to present worth. But now the task of
quantification is monumental. Nevertheless, systematic approaches
are described by many authors (see the collection of essays compiled
by [Layard and Glaister]).

Table 7.1.

Year	Savings	Costs	Present worth
0	0	− first costs	
1	$1,051.51	0	$1,030.69
2	$1,051.51	0	$1,010.28
3	$1,051.51	0	$990.27
4	$1,051.51	0	$970.67
5	$1,051.51	0	$951.45
6	$1,051.51	0	$932.61
7	$1,051.51	0	$914.14
8	$1,051.51	0	$896.04
9	$1,051.51	0	$878.29
10	$1,051.51	0	$860.90
11	$1,051.51	0	$843.86
12	$1,051.51	0	$827.15
	l.c. savings =	− first costs	+ $11,106.35

Exercises

7.1 Solve (7.2) for r.

7.2 Look at the lottery problem in §7.1 from the state's point of
view. Compute the purchase price q of an annuity that would
pay out $p/25$ per year for 25 years to the winner, given that the
remaining balance in the annuity earns at a continuous rate of
r per year.

7.3 What is the return on investment for the advanced component
heat pump of §7.2 assuming only $7000 in marginal first costs
and inflation $i = 3\%$?

Answer: 12.96%.

Outline: Using a spreadsheet, allocate one cell to contain the
putative discount rate d. Construct a table like Table 7.1, re-

ferring to the cell containing the discount rate d. Vary d until life cycle-savings are 0.

7.4 Redo the justified marginal first costs for the advanced heat pump of §7.2 when these first costs are borrowed at 5% continuously compounded yearly interest with a 15% down payment and a term of 12 years. (Use Exercise 3.6.)

Answer: $11,174.08.

Outline: Allocate one cell of a spreadsheet to contain the marginal first costs f. Construct a more detailed Table 7.1 that takes into account the initial payment and subsequent equal yearly payments determined from the assumed first costs f. Experiment with f until life-cycle savings fall to 0.

7.5 (Project) A solar system provides 67% of the domestic hot water needs of a household of four, each requiring 20 gallons of 140°F water per day, a rise of 90°F from water main temperature. Inflation is assumed to be 3% per year over the 15-year lifetime of the installation. The nonsolar load is met by an electric water heater; electric rates are $0.11 per kWh and rising at 0.15% per year over inflation. The cost of the solar system was $3600, with 20% returned from a state tax credit. The full cost $3600 was borrowed with no down payment at an interest rate of 6% over a term of 5 years. Assuming a discount rate of 5%, was this a good investment? If so, what is the return on investment and payback period? Display your analyses in tabular form. (See [Duffie and Beckman].)

7.6 (Project) Should all fossil-fueled generating plants be replaced by nuclear plants? Assume that a single uniform design for a nuclear plants has been chosen, to be duplicated without modification. (The present problems with the nuclear industry arise from the one-of-a-kind approach to construction.) Try to capture as many benefits and costs as possible, e.g., coal miner deaths, air pollution, future costs of spent fuel storage, expected cost of a meltdown and its environmental/health impact through the years, etc.

7.7 (Project) Should incandescent lights be turned off when leaving a room? Factor in the decreased life span from frequent turn-on surges versus energy savings. Is the answer the same for fluorescents? Formulate a typical usage pattern. Consider all costs and savings.

7.8 (Project) Is four years of study beyond the Master's to obtain a Ph.D. financially worthwhile? Does the added income throughout a career in industry pay back the cost of the education and loss of income during those 4 years?

7.9 Is deferring maintenance ever worthwhile?

7.10 (Project) Should class size be reduced nationwide in the public schools?

7.11 (Project) Should we colonize Mars?

7.12 (Project) Should passenger rail service be expanded in the northeastern corridor?

7.13 (Project) Should farmers be paid to plant winter cover crops? How can wind and water erosion of valuable topsoil be costed out?

7.14 (Project) Given the added selling price of organic produce, should a specialty farmer switch to organic methods?

Chapter 8
Microeconomics

Even the dismal science yields to mathematics.

Elementary economics provides good training in modeling. We look first at supply and demand, then revenue, cost, and profit, the notion of price elasticity, several models of competitive markets, a theory of production, and Leontiev's model of a national economy.

§8.1 Supply and Demand

Let us think about the market forces that control supply and demand for a single commodity. We assume that there are many producers and many consumers. The *demand* function

$$q = D(p) \tag{8.1}$$

gives the quantity q of units that will sell at price p, a decreasing function of p since higher price implies lower demand. Unfortunately, because of a historical misconception [Baldini et al., p. 6], demand curves are even today graphed with price as ordinate, as in Figure 8.1.

The *supply* function

$$q = S(p) \tag{8.2}$$

gives the number of units that all producers together will deliver at the price p. The supply function is, of course, an increasing function of price that is usually not positive before a minimum threshold is reached. A typical supply curve is displayed in Figure 8.1.

When demand and supply are graphed on the same axis as in Figure 8.1, their intersection at (q_0, p_0) yields up the *equilibrium price* p_0. Prices naturally seek this equilibrium: At prices above equilibrium, there are more goods being produced than the market will

absorb — this *surplus* drives down prices. Obversely, when prices
are below equilibrium, the demand is higher than the supply — this
scarcity will lift prices upward.

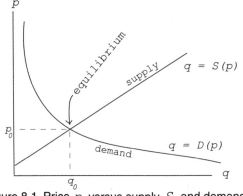

Figure 8.1. Price p versus supply S and demand D.
The two curves meet at the equilibrium (q_0, p_0).

Problem. What is the effect of a sales tax on price?

Solution. Since a sales tax is paid by the buyer, producers continue
to receive the price p per unit, so their decisions to supply remain
fixed — the supply curve $q = S(p)$ is unchanged. However, the buyer
must now pay $(1 + \epsilon)p$ per unit, which depresses demand, yielding
the new demand curve $q = D_\epsilon(p)$, given by

$$q = D_\epsilon(p) = D(p + \epsilon p), \tag{8.3}$$

as in Figure 8.2. Consequently, *sales taxes decrease equilibrium price
and quantity sold.* Moreover, producer's *revenue,*

$$R(p, q) = pq, \tag{8.4}$$

is decreased, while *public-sector revenue,*

$$G(p, q) = \epsilon pq, \tag{8.5}$$

is positive.

But note that as the sales tax rate ϵ is increased slowly, the
state's income (8.5) will at first rise, but as price and quantity fall,
its income may peak, then decrease. So it may not be in the interests
even of government to impose a tax beyond a certain critical rate.

Overall wealth may decrease (see Exercise 8.3). On the other hand, low tax rates may redistribute income in a socially undesirable way, underfunding infrastructure and education.

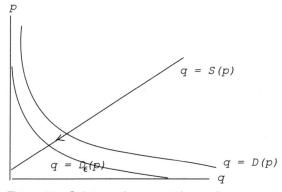

Figure 8.2. Sales tax depresses demand, moving equilibrium q and p downward.

§8.2 Revenue, Cost, and Profit

As just noted in (8.4), one particular producer's *revenue* $R(p, q)$ is the quantity sold times the price received for each unit sold. Assuming that actions of this one small producer cannot substantially affect the entire market, and in particular the equilibrium price p_0 remains constant, revenue for this producer becomes a function only of units sold:

$$R = R(q) = p_0 q. \tag{8.6}$$

This producer's *cost* to manufacture q units

$$C = C(q) \tag{8.7}$$

begins at some positive *fixed cost,* rises linearly at first, later flattens with economy of scale, then begins to rise at an ever-increasing rate as the producer becomes overwhelmed by the problems of high production: scheduling, labor shortages, overtime, etc. A typical cost curve is shown in Figure 8.3.

This producer's *profit* $\pi(q)$ *is revenue minus cost:*

$$\pi(q) = R(q) - C(q). \tag{8.8}$$

Question. Where should this business position itself? How many units should be produced?

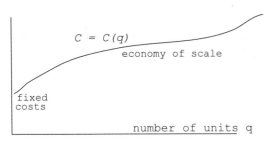

Figure 8.3. Typical cost function.

The answer in most situations is where profit is maximized, found by the following result.

RESULT A. As long as price remains constant, maximum (and minimum) profit occurs where marginal revenue equals marginal cost, i.e., when

$$R'(q) = C'(q), \tag{8.9}$$

i.e., where the curves of cost and revenue are parallel.

PROOF. Maximum profit occurs only where its derivative is zero.

REMARK. Economists use the word *marginal* to capture the notion of a derivative. For example, *marginal cost* is the additional amount (in dollars) required to produce one more unit of the commodity. Since quantities q are typically in the thousands or millions, this marginal change is for all practical uses the derivative with respect to q:

$$\frac{C(q+1) - C(q)}{1} \approx C'(q). \tag{8.10}$$

For another example, *marginal revenue* is the additional amount taken in by selling one more unit.

Example. Revenue from selling q (million) units is $R = 5q$, while cost is $C = 4 + q^2$. Thus revenue exceeds costs (and a profit is made) exactly when $5q > 4 + q^2$, i.e., when $1 < q < 4$. Maximum profit occurs when $\pi'(q) = 5 - 2q = 0$, i.e., when $q = 2.5$.

Should you make the 3,000,001th unit? No, since the marginal profit is negative — profit will decrease: $R'(3.000001) < C'((3.000001)$ (see Figure 8.4).

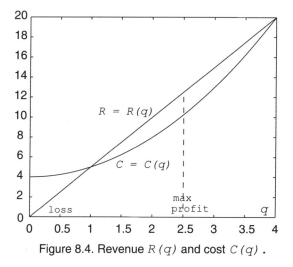

Figure 8.4. Revenue $R\,(q)$ and cost $C\,(q)$.

§8.3 Elasticity of Demand

Now step back to again view the entire market for one commodity. It is revealing to quantify how a price increase from p to $p + dp$ will affect demand $q = D(p)$. A statement like "raising our price $100 will decrease demand by 800 units" is useless out of context. If present demand is 900 units at a price of $1000 per unit, the result is disastrous. But if the present demand is 32,000, this price increase will be beneficial. A more revealing number would be the *percentage decrease in demand per percentage increase in price:*

$$e = -\frac{dq/q}{dp/p} = -\frac{dq}{dp} \cdot \frac{p}{q} = -p \cdot \frac{D'(p)}{D(p)}. \qquad (8.11)$$

This nonnegative dimensionless number e is called the *price elasticity of demand.* Demand is *elastic* if $e > 1$; i.e., small relative price increases have a relatively large effect on demand. Elastic demand is not good news. *Inelastic demand,* when $e \leq 1$, is a far better situation for the seller (see Result B below).

Example. In the examples mentioned above, a price increase of $100, up from $1000, with a corresponding drop of 800 from 900 units, is an elasticity $e \approx (800/900)/(100/1000) = 8.9$. Demand is elastic.

In the other case, a drop of 800 units out of 32,000 because of a price increase from \$1000 to \$1100 yields an elasticity $e \approx (800/32,000)/(100/1000) = 0.25$. Demand is inelastic.

RESULT B. Elastic demand means that revenue will fall with increasing prices. In fact,

$$\frac{dR}{dp} = D(p)(1 - e). \tag{8.12}$$

PROOF. Revenue $R = pq = pD(p)$, so

$$\frac{dR}{dp} = D(p) + pD'(p) = D(p)[1 + p\frac{D'(p)}{D(p)}]$$

$$= D(p)(1 - e). \tag{8.13}$$

Thus by (8.12), assuming costs remain parallel to revenue, *a monopoly will position itself on the demand curve at unitary elasticity ($e = 1$), thereby maximizing revenue.*

§8.4 Duopolistic Competition

Suppose that two competing comparable firms are the sole suppliers of essentially the same commodity — the traditional example is bottled water. We assume revenue and costs for each firm are parallel, so that profit for each is more or less independent of the number of units produced. Let us examine several models for the market dynamics arising from the competition between these two firms [Salvatore]. We will see that price, market share, and revenue are strongly affected by how each firm views its competitor. Both producers together are subject to the demand curve $q = D(p)$.

Example. (The Cournot model, 1838) Each firm believes that the other will hold its output constant.

In the beginning, firm A dominates the market completely and is positioned at the monopolistic price $p = a_1$ determined by (8.12), i.e., where $e = 1$. Firm B now enters the market by selling at a lower price. Since firm B believes that firm A will hold its production at $q_{A1} = D(a_1)$ units, the demand curve available for firm B is demand $q = D(q)$ translated downward to the remaining market:

$$D_{B1}(p) = D(p) - q_{A1}. \tag{8.14}$$

Firm B now positions itself at price $p = b_1$ determined by (8.13) for maximum revenue.

Firm A next responds to its perceived new demand curve,

$$D_{A1}(p) = D(p) - q_{B1}, \tag{8.15}$$

where $q_{B1} = D_{B1}(b_1)$, and lowers its price to $p = a_2$ according to (8.12). These actions and responses yield the discrete-time system

$$a_n D'(a_n) + D(a_n) = -b_n D'(b_n) \tag{8.16a}$$

$$b_n D'(b_n) + D(b_n) = -a_{n-1} D'(a_{n-1}) \tag{8.16b}$$

(Exercise 8.8). Generically, we may invert the functions involved to obtain the system

$$a_n = F(b_n) \tag{8.17a}$$

$$b_n = F(a_{n-1}) \tag{8.17b}$$

so

$$a_n = F(F(a_{n-1})) \tag{8.18a}$$

$$b_n = F(F(b_{n-2})). \tag{8.18b}$$

So by the fixed-point theorem [Franklin], supposing F contractive, prices of both firms eventually converge to the identical *equilibrium* intermediate price p_∞ (see Exercises 8.9 and 8.10). The revenue of each firm is typically less than if they cooperatively divided the market in half.

Example. (The Bertrand model, 1883) Each firm believes that the other will hold its price constant. The firm with the lower price captures the entire market.

Then each firm will lower its price below the other firm's price until price vanishes and revenue ceases. A silly model.

Example. (The Edgeworth model, 1925) Each firm believes the other will hold its price constant. Customers will opt for the unit of lower price when available. Both firms have a limited output.

As in the Bertrand model, a price war ensues, with prices dropping and output increasing until both firms reach and surpass maximum production q_A and q_B. The price war continues with prices dropping even further, to p_{lo}, until one firm, say A, realizes that there is a much higher price p_{hi} where leftover demand from q_B yields more revenue than does its present price-war position p_{lo}. Firm A moves to this higher price p_{hi}. Firm B sees this and raises prices to just below p_{hi}, thus reigniting the price war. The system oscillates. You are asked to formalize this model in Exercise 8.12.

Duopolistic competition is mathematically fertile (see [Puu]).

§8.5 Theory of Production

Production is a function of labor and capital investment. The most common model is the *Cobb–Douglas*[1] model [Cobb and Douglas]:

$$Q = AL^\alpha K^{1-\alpha}, \tag{8.19}$$

where Q is the number of units produced, L the number of workers needed for this output, K the capital expended for this output, and A is a factor (with appropriate units) arising from the technology employed to produce the good. Note that $Q = Q(L, K)$ is homogeneous of degree 1, i.e., $Q(\lambda L, \lambda K) = \lambda Q(L, K)$, so there is *constant return to scale* — doubling labor and capital doubles the output. Moreover, by Euler's identity (Exercise 8.13),

$$Q = L\frac{\partial Q}{\partial L} + K\frac{\partial Q}{\partial K}. \tag{8.20}$$

From the "neoclassic" economic view, the wage w paid to a worker should equal the *marginal product of labor* $\partial Q/\partial L$, the additional output from one additional worker:

$$w = \frac{\partial Q}{\partial L} = \frac{\alpha Q}{L}, \tag{8.21}$$

so that workers are paid at the same rate they produce. On the other hand, firms should only pay rent r for capital (machinery, building space, etc.) equal to the *marginal product of capital*:

$$r = \frac{\partial Q}{\partial K} = \frac{(1-\alpha)Q}{K}. \tag{8.22}$$

Combining (8.20)–(8.22) gives

$$Q = wL + rK = \alpha Q + (1 - \alpha)Q. \tag{8.23}$$

---

[1]Paul Douglas, U.S. Senator from Illinois, 1948–1966.

Equation (8.23) is also accepted as macroeconomic law, modeling the total market of entire countries [Cobb and Douglas]. For example, the percentage contribution of labor $\alpha = Q_L/Q$ in the United States is estimated to be $\alpha = 0.67$, while the contribution of capital is 0.33 [Klein; Abel and Bernayke, p. 62].

For an individual firm, the Cobb–Douglas model (8.19) can be used for decisions on whether to hire or to spend on capital improvements.

Example. Suppose that a firm is producing 100 units with a labor force of 30 and capital investment of 40 (in appropriately sized units). Let us assume the labor contribution to be $\alpha = 0.65$. Then $100 = A \cdot 30^{0.65} \cdot 40^{0.35}$, so $A = 3.01$. At this operating point, according to (8.21) and (8.22), a worker's wage should be $w = \alpha Q/L = 2.17$ and rent $r = (1-\alpha)Q/K = 0.88$. Thus production Q is governed nearby this present operating point by

$$Q = 2.17L + 0.88K. \qquad (8.24)$$

So at first glance, output is far more sensitive to labor L than to capital investment. But there may be constraints. There may be only so many machines available — hiring more only means idle workers waiting for machine free time. There may be many other constraints. This is a linear programming problem in the two variables L and K subject to possibly many constraints (see Exercise 8.13).

§8.6 Leontiev Input/Output

The following macroeconomic model is included because of its mathematical charm. Imagine a simplified national economy with only three producing sectors — agriculture, industry, and service — together with the nonproducing consumer sector, as tabulated in Table 8.1. The first row of table depicts how agriculture's total production of x_1 units is consumed: 30% is consumed by agriculture itself, $0.2x_2$ is used by industry as it produces x_2 units, $0.3x_3$ is used by the service industry as it produces x_3 units, and the remaining 4 units are absorbed by the consumer sector. In short, the basic assumption is that

the jth sector input of the ith sector's output x_i is proportional to the jth sector's output x_j.

Table 8.1

	Ag.	Indust.	Serv.	Consumer	Total prod.
Ag.	$0.3x_1$	$0.2x_2$	$0.3x_3$	4	x_1
Indust.	$0.2x_1$	$0.4x_2$	$0.3x_3$	5	x_2
Serv.	$0.2x_1$	$0.5x_2$	$0.1x_3$	12	x_3
				(bill of goods)	

Rewrite Table 8.1 as the matrix relation

$$Ax + b = x. \tag{8.25}$$

Question. Does the vector equation (8.25) have a nonnegative solution $x \geq 0$? Is the *bill of goods* obtainable?

The bill of goods is certainly obtainable when $I - A$ has an inverse with all entries nonnegative, for then

$$x = (I - A)^{-1} b \geq 0. \tag{8.26}$$

RESULT C. If the dominant eigenvalue of A is less than 1, every (nonnegative) bill of goods b is realizable by a (nonnegative) production output x.

PROOF SKETCH. Every matrix is asymptotically diagonal — a slight modification of the proof of the Jordan canonical form will bring a matrix to a similar matrix that is within ϵ of a diagonal matrix. So if all eigenvalues of A are less than 1 in modulus, we may use the similarity to define a norm where $\|A\| < 1$. But then the geometric series

$$I + A + A^2 + A^3 + \cdots = (I - A)^{-1}$$

converges to a matrix with nonnegative entries.

En passant, the Frobenius–Perron theorem guarantees that a dominant eigenvalue λ of a nonnegative matrix A is nonnegative and belongs to a nonnegative eigenvector.

Leontiev's models — both static and dynamic — are discussed at length in [Dorfman et al.].

Exercises

8.1 Demand is $D(p) = 2/p$ while supply is $S(p) = -1+p$. Find the equilibrium price p (in dollars) and quantity q (in thousands of units).

8.2 Find the equilibrium price and quantity in Exercise 8.1 after instituting a sales tax on the consumer of $\epsilon = 5\%$. What is the producer's revenue? What is the state's revenue?

8.3 Formulate a concept of *total wealth*. Has total wealth increased or decreased in Exercise 8.2 after the sales tax was imposed?

8.4 Given that revenue $R(p, q) = pq = qD^{-1}(q)$, show that marginal revenue $dR/dq = p(1 - 1/e)$.

8.5 Show that the demand curve $D(p) = 1/p$ has a coefficient of elasticity identically 1; i.e., revenue is constant with price. Find all such curves.

8.6 Show that for linear demand curves $D(p) = q_0 - mp$, the coefficient of elasticity e is 1 only at midpoint, i.e., at $p = q_0/2m$.

8.7 Find a demand curve with no price of unitary elasticity $e = 1$. Find another demand curve with exactly two such prices.

8.8 Verify that the recursion (8.16) captures Cournot competition.

8.9 Verify graphically and analytically that the Cournot model of duopolistic competition (8.16) converges for a linear demand function $D(p) = q_0 - mp$. Show that the final revenue of each firm is less than if they had formed a *cartel* and divided the market in half.

Answer: $p_\infty = q_0/3m$.

8.10*(Project) Characterize the demand functions $q = D(p)$ for which the Cournot model of duopolistic competition converges. Also decide which demand functions $q = D(p) - q_0$ have a unique price of maximal revenue for all reasonable $q_0 > 0$.

8.11 Revenue appears as what geometric object superimposed upon the graph of price against demand $q = D(p)$?

8.12 (**Project**) Formalize the Edgewood oscillating duopolistic competition model of §8.4 as a discrete-time system. Find the price rails p_{lo} and p_{hi} for a linear demand curve $D(p) = q_0 - mp$ and production limits q_A and q_B. Find an expression for these price rails for a general demand function. What mathematical assumptions on the demand function are necessary to make this model rigorous?

8.13 Suppose that F is homogeneous of degree m, i.e.,

$$F(\lambda x) = \lambda^m F(x),$$

where $x = (x_1, x_2, \dots, x_n)$. Show that

$$mF(x) = x_1 F_{x_1}(x) + x_2 F_{x_2}(x) + \cdots + x_n F_{x_n}(x).$$

8.14 Extend the Cobb–Douglas example of §8.5 by proposing at least four constraints, then solve the resulting linear programming program.

8.15 Is the bill of goods shown in Table 8.1 obtainable?

8.16 Suppose the producer is required to pay a *specific* tax of $1.00 per unit (as with fuel, liquor, or tobacco). If the before-tax demand and supply were $D(p) = 100 - 30p$ and $S(p) = 20p - 50$, what are the before- and after-tax equilibrium prices? How much of the tax per unit is actually paid by the consumer and how much by the producer?

Answers: $3, $3.4, $0.40, $0.60.

8.17 In contrast to Exercise 8.16, rather than a specific tax, suppose the producer is required to pay a sales tax of 33%. What are the before- and after-tax equilibrium prices? How much of the tax per unit is actually paid by the consumer and how much by the producer?

Chapter 9

Ordinary Differential Equations

Differential equations enable us to reconstruct the past and predict future events. They are unreasonably effective in their power to model the world about us.

We apply the method of separation of variables in two disaster scenarios and to population growth, each of the three solutions illustrating a fundamental approach to modeling. Mechanical models are obtained by force balancing and by energy conservation methods. Heaviside's 'big D' method is reviewed and applied to problems in motion and electronic resonance. Systems of ODEs are obtained from modeling heat exchangers, damped harmonic motion, pursuit, a stochastic evolution, and expressway traffic.

§9.1 Separation of Variables

Recall the first method of this subject: If an ordinary differential equation

$$\frac{dy}{dx} = F(x, y) \tag{9.1}$$

is of the special form

$$\frac{dy}{dx} = f(x)g(y), \tag{9.2}$$

then we *separate variables*

$$\frac{dy}{g(y)} = f(x)\, dx, \tag{9.3}$$

and integrate to obtain the often implicit solution

$$\int \frac{dy}{g(y)} = \int f(x)\, dx + C. \tag{9.4}$$

Example. Separating variables and integrating the ODE $y' = x/y$ yields

$$\frac{y^2}{2} = \int y\, dy = \int x\, dx = \frac{x^2}{2} + C,$$

i.e.,

$$y^2 = x^2 + 2C. \tag{9.5}$$

Picard's fundamental theorem [Boyce and DiPrima, p. 41] guarantees that among this family of solutions there is exactly one local solution $y = y(x)$ satisfying, say, the *boundary/initial condition* $y(3) = 5$. But then from (9.5), $5^2 = 3^2 + 2C$, so $C = 8$. This unique solution $y = y(x)$ satisfies (9.5) and thus must be the positive branch

$$y = \sqrt{x^2 + 16} \tag{9.6}$$

since the negative branch $y = -\sqrt{x^2 + 16}$ does not meet the boundary condition $y(3) = 5$.

Problem. An attack submarine is cruising at 40 knots at a depth of 1000 feet when suddenly the reactor scrams. After 1 minute way has dropped to 30 knots. How long does the crew have to make repairs before forward motion falls below steerageway of 2 knots?

Solution. At first thought the craft is losing 10 knots per minute, so will come to rest after 4 minutes. But the situation is not nearly so dire. As the craft slows, water resistance decreases, and thus momentum is bleeding off at a slower and slower rate. Intuitively, it appears as if the craft will never come to rest.

There is a simple and persuasive argument in Exercise 9.1 that the (force of) water resistance against such a craft is more or less proportional to the square of its velocity, i.e.,

$$R = -kv^2. \tag{9.7}$$

(The negative sign reflects that the resistance is in opposition to the velocity.) We now apply *Newton's rule,*

$$\sum inertial\, forces = \sum external\, forces. \tag{9.8}$$

The inertial forces are one in number, namely $F = ma = m\dot{v}$, where m is the mass of the submarine. The external forces are also one in number, viz, $R = -kv^2$. Equating, we obtain our model:

$$m\frac{dv}{dt} = -kv^2 \tag{9.9}$$

subject to the data

t	v
0	40
1	30
?	2

Lump the unknown constants into one: $\dot{v} = -Kv^2$, separate variables, and integrate to obtain

$$\frac{1}{v} = Kt + C. \tag{9.10}$$

Using our initial datum $v(0) = 40$, we see that $C = 1/40$, i.e.,

$$\frac{1}{v} = Kt + \frac{1}{40}. \tag{9.11}$$

Using our second datum $v(1) = 30$, we see that $K = 1/30 - 1/40 = 1/120$. All future velocities are thus given by

$$\frac{1}{v} = \frac{t}{120} + \frac{1}{40}. \tag{9.12}$$

Placing steerageway $v = 2$ into (9.12) yields $t = 57$, i.e., 56 minutes remain before steerageway is lost.

Alternatively, the model (9.9) can be deduced by equating the rate of loss of kinetic energy $d(mv^2/2)/dt$ to the rate Rv that work is being done against the water.

Observe how the submarine problem above tracks exactly what some wags call the "scientific method":

Step 1. Model the phenomenon as a differential equation.

Step 2. Solve the differential equation.

Step 3. Impose the given data.

Step 4. Interpret the results.

Problem. A house furnace fails on a cold winter's evening when the outside (ambient) temperature $T_a = 20°F$. Although initially at $70°F$, the inside temperature T has fallen to $65°F$ after 1 hour. How long before the inside temperature reaches the damaging temperature of $32°F$?

Solution. We employ Polya's rule: *Derive some quantity in two different ways and equate*[Polya]. In this case let us derive the instantaneous rate $\dot{Q}$ at which heat is being lost from the house to ambient. On the one hand, the rate at which heat is driven through insulation (by conduction) is proportional to the temperature gradient driving out that heat, i.e.,

$$\dot{Q} = U(T - T_a), \tag{9.13}$$

just as current I is driven through a resistance R by the potential V across it: $I = V/R$ (Ohm's law).

On the other hand, where is this heat coming from? It is released from the furnishings and objects within the house, from the thermal mass within the envelope. Heat is the energy released when objects are lowered in temperature. The rate that heat is being liberated from the interior is proportional to the rate at which the interior is dropping in temperature, i.e.,

$$\dot{Q} = -M\dot{T}. \tag{9.14}$$

By equating (9.13) and (9.14), we have our model:

$$\dot{T} = -\frac{U}{M}(T - T_a) \tag{9.15}$$

subject to the data

t	T
0	70
1	65
?	32

This common model (9.15) is known as *Newton's Law of Cooling*. Solve by separating variables to obtain (Exercise 9.2) that interior temperature $T(t)$ drops below 32°F when $t \approx 13.5$ hours.

Practice Newton's law of cooling by solving the modeling Exercises 9.3 through 9.6.

Problem. Predict the future population of a developed country.

Solution. The timerate of increase $\dot{P}$ of the population (in millions of people/year) is at first cut proportional to the number of people P; doubling the number of reproducing adults should double the number of births per year. This gives the model $\dot{P} = kP$ with exponential solution $P = P_0 e^{kt}$. Such exponential growth would have outstripped resources long ago. This model cannot be correct.

Apparently there are factors inhibiting such exponential growth that become more dominant as a population grows. The model must be of a more complex form

$$\dot{P} = f(P).$$

Try the first two terms of the Taylor series of $f(x)$:

$$\dot{P} = kP - \epsilon P^2. \tag{9.16}$$

Surprisingly this model, called the *logistics equation*, fits available data moderately well. At first the linear term kP dominates yielding exponential-like growth. But as population builds, the quadratic term ϵP^2 will come into play to inhibit growth rate. Equilibrium is reached if growth rate ceases: $\dot{P} = 0 = kP - \epsilon P^2$, i.e., at population $P_\infty = k/\epsilon$. This upper population limit P_∞ is called the *carrying capacity* of the society. However the carrying capacity, itself a constant solution to the logistics equation (9.16), will never be achieved in finite time since — as a byproduct of the uniqueness of local solutions guaranteed by Picard's theorem —

solutions cannot meet.

Population modeling is a profession in its own right — see Hoppensteadt's entertaining *Mathematical Methods of Population Biology* or the classic [Keyfitz and Flieger].

§9.2 Mechanics

There are two elementary approaches to modeling mechanical systems. One I call *Newton's method,* where inertial forces are equated with external forces, the other *Hamilton's method,* where total energy is assumed constant along a trajectory of motion. Both methods are deductions from the *Lagrange equations* [Goldstein].

Example. (Planar pendulum) Suspend a mass m by a rigid massless rod of length L. The motion of this pendulum is constrained to lie in one fixed plane. Thus this simple machine has exactly one *degree of* *freedom* — it is determined completely by the one angular variable θ as shown in Figure 9.1.

Newton's Method: The tangential inertial force $F = ma$, where $a = L\ddot{\theta}$. The only external force is due to gravity with tangential component $-mg\sin\theta$. Equating tangential forces gives the model

$$\ddot{\theta} = -\frac{g}{L}\sin\theta. \tag{9.17}$$

Hamilton's Method: Kinetic energy is $T = mv^2/2 = m(L\dot{\theta})^2/2$, while potential energy V is the work done lifting m vertically against gravity from bottom dead center up to its present height; i.e., $V =$ force $\cdot$ distance $= mg \cdot (L - L\cos\theta)$. Assuming that no energy is being bled off by friction, any initial energy E_0 must be conserved as the motion evolves; i.e., $T + V = E_0$ for all $t \geq 0$. Thus

$$\frac{m(L\dot{\theta})^2}{2} + mg(L - L\cos\theta) = E_0,$$

i.e.,

$$\frac{\dot{\theta}^2}{2} + \frac{g}{L}(1 - \cos\theta) = \frac{E_0}{mL^2}. \tag{9.18}$$

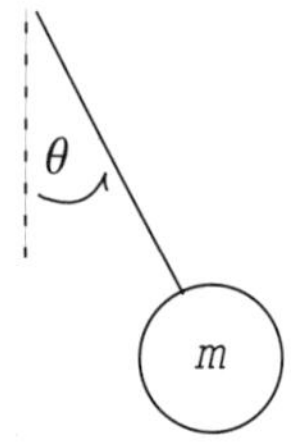

Figure 9.1. Planar pendulum.

Note that differentiating Hamilton's model (9.18) yields Newton's model (9.17).

Example. (Mass–spring) The mass m is sliding along a frictionless horizontal table restrained by a spring with spring constant k (Figure 9.2). Motion is assumed linear. The displacement $x = x(t)$ of the mass m is measured from equilibrium (rest).

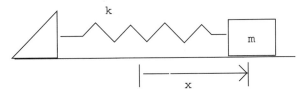

Figure 9.2. Spring-mass system

Newton's Method: Inertial force $F = ma = m\ddot{x}$. The only external force is the restoring force of the spring $-kx$. Thus

$$m\ddot{x} + kx = 0. \tag{9.19}$$

Hamilton's Method: Kinetic energy is $T = mv^2/2 = m\dot{x}^2/2$ while the potential energy V is the work done moving m against the force of the spring from equilibrium to its present position x. Thus

$$\frac{m\dot{x}^2}{2} + \int_0^x ky\,dy = E_0,$$

i.e.,

$$\frac{m\dot{x}^2}{2} + \frac{kx^2}{2} = E_0. \tag{9.20}$$

Again note that the derivative of Hamilton's solution is Newton's solution. Alternatively, multiply Newton's model through by $\dot{x}$ and integrate to obtain Hamilton's model.

Example. (Rolling spring–mass) A mass m of radius r and moment of inertia I is rolling without slippage along a horizontal table restrained by a spring of constant k (see Figure 9.3). Motion is along a straight line, and displacement x is measured from equilibrium.

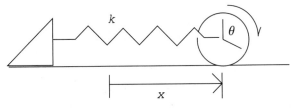

Figure 9.3. Rolling spring-mass.

Newton's Method: The inertial force is the sum of the linear and rotational forces: $F + \tau/r = m\ddot{x} + I\ddot{\theta}/r = m\ddot{x} + I\ddot{x}/r^2$, while the external restoring force is $-kx$. Thus

$$m\ddot{x} + \frac{I\ddot{x}}{r^2} = -kx.$$

REMARK. A more precise approach would be to balance translational forces and rotational torques separately to yield a coupled system:

$$m\ddot{x} = -kx - \frac{\tau}{r}$$

$$I\ddot{\theta} = \tau.$$

Hamilton's Method: The sum of linear and rotational kinetic energy is $T = mv^2/2 + I\dot{\theta}^2/2 = m\dot{x}^2/2 + I(\dot{x}/r)^2/2$, while potential is the work done stretching the spring: $V = kx^2/2$. Thus

$$\frac{m\dot{x}^2}{2} + \frac{I(\dot{x}/r)^2}{2} + \frac{kx^2}{2} = E_0.$$

You may review these notions in the very readable Schaum's Outline *Applied Physics* [Beiser]. At the other end is Goldstein's *Classical Mechanics*.

§9.3 Linear ODEs with Constant Coefficients

An astonishing number of phenomena are well modeled by linear ODEs with constant coefficients,

$$y^{(n)} + a_{n-1}\, y^{(n-1)} + a_{n-2}\, y^{(n-2)} + \cdots + a_1\, y' + a_0\, y = u(x), \quad (9.21)$$

where the a_i are fixed real numbers. The function $u(x)$ on the right of (9.21) is called the *driving function* or the *input*.

Example. The basic spring–mass system shown in Figure 9.2 is modeled by equating the inertial force $ma = m\ddot{x}$ to the restoring force of the spring:

$$m\ddot{x} = -k(x - x_0) \quad \text{(harmonic motion)}, \quad (9.22)$$

where x_0 is the displacement x of the mass m when in equilibrium (at rest). If there is a second external force of *viscous friction,* i.e., friction proportional to velocity as in Figure 9.4, the model becomes

$$\ddot{x} + c\dot{x} + kx = kx_0.$$

Note that by measuring from equilibrium, by setting $y = x - x_0$, this driven system becomes the *freely running, homogeneous* model

$$\ddot{y} + c\dot{y} + ky = 0 \quad \text{(damped harmonic motion)}. \tag{9.23}$$

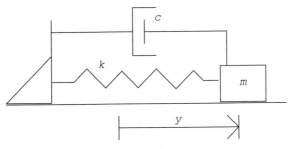

Figure 9.4. Damped spring-mass system.

Example. You will learn in Chapter 10 how to model linear electronic circuits. In particular, consider the ubiquitous frequency-determining parallel resonant circuits shown in Figure 9.5, found in any radio or television set. As you will see, the input u and output voltages y are related by

$$\ddot{y} + \frac{\dot{y}}{RC} + \frac{y}{LC} = \frac{\dot{u}}{RC}, \tag{9.24}$$

so that if the input is grounded at $t = 0$, i.e., $u(t) = 0$ for $t \geq 0$, the voltage y satisfies damped harmonic motion (9.23) with $c = 1/RC$ and $k = 1/LC$.

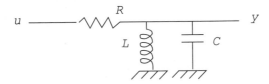

Figure 9.5. Parallel resonant circuit.

As we see a few lines below, increasing damping c lowers vibrational frequency and can even extinguish vibration altogether. You may have noticed the large amount of damping in the front suspensions of high performance automobiles. Depressing, then releasing the front bumper will yield no vibration nor even any overshoot (in stark contrast to older passenger vehicles with lighter damping).

Homogeneous constant coefficient linear equations are solved by *Heaviside's 'big-D' method:*

THEOREM A. The general solution on $(-\infty, \infty)$ of the homogeneous linear ODE with constant coefficients

$$y^{(n)} + a_{n-1}\, y^{(n-1)} + a_{n-2}\, y^{(n-2)} + \cdots + a_1\, y' + a_0\, y = 0 \qquad (9.25)$$

is obtained by the following algorithm:

Step 1. Write the ODE (9.25) in operator format:

$$p(D)y = (D^n + a_{n-1}D^{n-1} + \cdots + a_1 D + a_0)y = 0.$$

Step 2. Factor the operator (β_k distinct)

$$p(D) = (D - \beta_1)^{m_1}(D - \beta_2)^{m_2} \cdots (D - \beta_r)^{m_r}.$$

Step 3. Write down the general solution

$$y = \sum_{k=1}^{r} e^{\beta_k x} \sum_{j=0}^{m_k - 1} c_{jk} x^j. \qquad (9.26)$$

Example. The ODE $y'' + 3y' + 2y = 0$ becomes in operator format

$$(D^2 + 3D + 2)y = 0,$$

i.e., when factored

$$(D + 1)(D + 2)y = 0$$

with general solution

$$y = c_1 e^{-x} + c_2 e^{-2x}.$$

Example. The ODE $y'' + 2y' + y = 0$ becomes

$$(D^2 + 2D + 1)y = (D + 1)^2 y = 0,$$

with general solution

$$y = (c_0 + c_1 x)e^{-x}.$$

Example. Undamped harmonic motion $\ddot{y} + ky = 0$ becomes

$$(D^2 + k)y = (D - i\sqrt{k})(D + i\sqrt{k})y = 0,$$

with general solution

$$
\begin{aligned}
y &= c_1 e^{i\sqrt{k}t} + c_2 e^{-i\sqrt{k}t} \\
&= (\text{Exercise 9.10}) = a \cos \sqrt{k}t + b \sin \sqrt{k}t.
\end{aligned}
\tag{9.27}
$$

Example. Damped harmonic motion $\ddot{y} + c\dot{y} + ky = 0$ rewritten in standard engineering notation becomes

$$\ddot{y} + 2\zeta\omega_n\dot{y} + \omega_n^2 y = 0, \tag{9.28}$$

where ω_n is the *undamped natural frequency* of oscillation and where ζ is the *damping ratio*. These terms are deserved, as we now see.
Write (9.28) in operator format:

$$[D - (-\zeta\omega_n + \sqrt{\zeta^2 - 1}\omega_n)][D - (-\zeta\omega_n - \sqrt{\zeta^2 - 1}\omega_n)]y = 0. \tag{9.29}$$

There are three cases.

Case 1. The system is underdamped, i.e., $\zeta < 1$.
    In this case roots are complex and vibration occurs:

$$
\begin{aligned}
y = \cdots &= e^{-\zeta\omega_n t}(a \cos \omega_n \sqrt{1 - \zeta^2}t + b \sin \omega_n \sqrt{1 - \zeta^2}t) \\
&= ce^{-\alpha t} \cos(\omega_d t + \phi)
\end{aligned}
\tag{9.30a}
$$

where $\alpha = \zeta\omega_n$ is called the *damping coefficient* while $\tau = 1/\alpha$ is the *time constant* of the system. Note that as the damping ratio ζ increases from 0, *damped vibrational frequency* $\omega_d = \omega_n\sqrt{1 - \zeta^2}$ decreases. Vibration is extinguished when ζ reaches 1.

Case 2. The system is critically damped, i.e., $\zeta = 1$.
    In this case the roots are real and repeated, giving the eventually decaying response

$$y = (c_0 + c_1 t)e^{-\alpha t}. \tag{9.30b}$$

Case 3. The system is overdamped, i.e., $\zeta > 1$.

In this final case there are two distinct roots, both negative, so the solution is the sum of two exponentially decaying functions:

$$y = c_1 e^{-(\zeta - \sqrt{\zeta^2 - 1})\omega_n t} + c_2 e^{-(\zeta + \sqrt{\zeta^2 - 1})\omega_n t}. \tag{9.30c}$$

Example. The third-order problem $y''' + 2y'' - y' - 2y = 0$ becomes $(D - 1)(D + 1)(D + 2)y = 0$ with general solution

$$y = c_1 e^x + c_2 e^{-x} + c_3 e^{-2x}$$

that is unbounded as $x \to \infty$ unless $c_1 = 0$. Positive roots signal instability.

Example. The seventh-order equation $(D + 2)^3 (D^2 + 1)^2 y = 0$ has general solution

$$y = e^{-2t}(c_0 + c_1 t + c_2 t^2) + (a_0 \cos t + b_0 \sin t) + t(a_1 \cos t + b_1 \sin t).$$

Problem. We wish to travel from point A to point B on the surface of the Earth by falling (or rolling) down the tunnel bored along the chord connecting A to B. How long will the trip take?

Solution. The acceleration due to gravity changes from an inverse square law $a = -GM/r^2$ off the planet to a proportional rule $a = -kr$ within the crust (see Exercise 9.12). Let $x = x(t)$ be the distance of our rolling vehicle from the midpoint of the chord AB as shown in Figure 9.6. By similarity,

$$\frac{-\ddot{x}}{kr} = \frac{x}{r},$$

and thus travel is governed by harmonic motion

$$\ddot{x} = -kx$$

of frequency $\omega = \sqrt{k}$ of period $T = 2\pi/\omega$. Thus the duration of the trip $T/2$ is the same regardless of the length of the journey!

At the surface, $g = 32/5280 = kR$ (miles/s^2), where $R = 3964$ miles (Exercise 9.14), giving that the half period $T/2 = \pi/\sqrt{k} = 2540.73$ seconds $= 42$ minutes and 21 seconds.

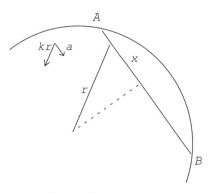

Figure 9.6. Tunnel bored along chord.

§9.4 Systems

It is modern and effective to trade order for dimension by going over to *state space*.

Example. The third-order linear ODE $y''' + 3xy'' - e^x y' - 5y = \sin x$ becomes, after we introduce the *states* $y_1 = y$, $y_2 = y'$, $y_3 = y''$, the first-order system

$$
\begin{aligned}
y_1' &= & y_2 & & & \\
y_2' &= & & y_3 & & \\
y_3' &= & 5y_1 &+ e^x y_2 & - 3xy_3 & + \sin x.
\end{aligned}
$$

Put in matrix notation,

$$
\begin{vmatrix} y_1 \\ y_2 \\ y_3 \end{vmatrix}' = \begin{pmatrix} 0 & 1 & 0 \\ 0 & 0 & 1 \\ 5 & e^x & -3x \end{pmatrix} \begin{vmatrix} y_1 \\ y_2 \\ y_3 \end{vmatrix} + \begin{vmatrix} 0 \\ 0 \\ \sin x \end{vmatrix}. \tag{9.31}
$$

In short the state space system is of the form

$$
y' = Ay + u.
$$

Mechanics problems are often examined in *phase space*:

Example. Damped harmonic motion $m\ddot{y} + c\dot{y} + ky = 0$ of Figure 9.4 with the traditional assignment $q = y$ (displacement) and $p = m\dot{y}$ (momentum) becomes the first-order system

$$\dot{q} = \qquad\qquad p/m$$
$$\dot{p} = \quad -kq \quad -cp/m \quad.$$

In matrix notation,

$$\left|\begin{array}{c}\dot{q}\\\dot{p}\end{array}\right| = \left(\begin{array}{cc}0 & 1/m\\-k & -c/m\end{array}\right)\left|\begin{array}{c}q\\p\end{array}\right|. \qquad (9.32)$$

Example. The multiple mass–spring system (Figure 9.7), where x_i is the displacement of the ith mass m_i from equilibrium, satisfies the second-order system

$$m_i\ddot{x}_i = k_{i-1}x_{i-1} - (k_{i-1} + k_i)x_i + k_i x_{i+1} \qquad (9.33\text{a})$$

$i = 1, 2, \ldots, n$ (Exercise 9.23). In matrix form,

$$M\ddot{x} + Kx = 0, \qquad (9.33\text{b})$$

where M is the diagonal *mass matrix* and K is the positive definite symmetric banded tridiagonal matrix of *stiffness* (Exercise 9.25).

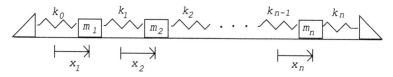

Figure 9.7. Multiple mass-spring system.

Problem. A drake spies a duck on the bank and begins swimming toward her. Model his path.

Solution. Let us assume that the river current is more or less of constant (vector) velocity $\mathbf{v_c}$ and that the drake maintains a constant water speed of v_d in the direction from his present position $\mathbf{r}(t) = (x(t), y(t))$ toward her position $\mathbf{b}$. Then his velocity with respect to a fixed coordinate system is the sum of the river current and his own efforts:

$$\dot{\mathbf{r}} = \mathbf{v_c} + v_d\frac{\mathbf{b} - \mathbf{r})}{|\mathbf{b} - \mathbf{r}|}. \qquad (9.34)$$

Establishing the coordinate system as shown in Figure 9.8, our vector model (9.34) becomes the system

$$\dot{x} = -v_c - \frac{v_d x}{\sqrt{x^2 + (b - y)^2}} \qquad (9.35\text{a})$$

$$\dot{y} = \frac{v_d(b-y)}{\sqrt{x^2 + (b-y)^2}}. \qquad (9.35b)$$

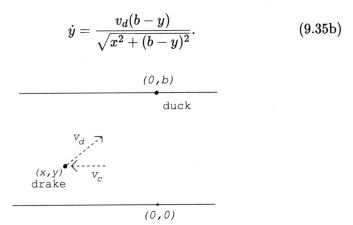

Figure 9.8. A drake paddles toward a duck at $(0,b)$.

If you need only the path and not its time dependence, there is this classic trick:

$$\frac{dy}{dx} = \frac{dy/dt}{dx/dt} = \frac{-v_d(b-y)}{v_c\sqrt{x^2 + (b-y)^2} + v_d x}. \qquad (9.36)$$

Problem. A *tube-within-tube counterflow* heat exchanger is carrying two fluids of temperatures $t(x)$ and $T(x)$ as in Figure 9.9. Given the four entering and exiting temperatures, predict both fluid temperatures within the tube.

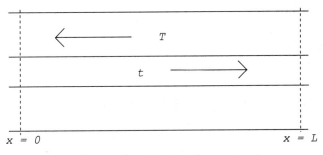

Figure 9.9. Counterflow tube-in-tube heat exchanger.

Solution. We do the case where the outer tube is insulated from surrounding temperatures. Assume that heat exchange is indeed taking place, that $t(0) \neq T(L)$. Two applications of Exercise 9.4 yield the system

$$t' = -k(t - T) \qquad (9.37a)$$

$$T' = -K(T - t) \tag{9.37b}$$

(see Exercise 9.4). Let us first solve for the gradient $\theta = t - T$. From (9.37), $\theta' = -(k + K)\theta = -\alpha\theta$, so $\theta = ce^{-\alpha x}$. Setting $x = 0$, we see that $c = t(0) - T(0)$. Setting $x = L$ yields

$$\alpha = \ln \left[\frac{t(0) - T(0)}{t(L) - T(L)} \right]^{1/L}.$$

But then integrating $t' = -k\theta$ and $T' = K\theta$ yields our prediction

$$t = be^{-\alpha x} + d \tag{9.38a}$$

$$T = Be^{-\alpha x} + D, \tag{9.38b}$$

where α is given above and where the four constants b, d, B, and D can be determined from the four given entering and exiting temperatures $t(0), T(0), t(L)$, and $T(L)$ (Exercise 9.26).

Problem. Every day you must give your dog one half of a heartworm pill. You begin the mosquito season with a large bottle of N whole pills. Each day you decant an object from the bottle: If it is a half pill, give it to the dog; if whole, break it in half, administer one half to the dog, and return the other half to the bottle. After k days, what is the number of unbroken pills remaining? (This is a reprise of Exercise 2.17.)

Solution. Let x_k and y_k be the expected number of whole and half pills remaining after the kth ministration, respectively. At the $(k + 1)$th extraction, the expected net change in whole pills will be

$$x_{k+1} - x_k = -\frac{x_k}{x_k + y_k} \tag{9.39a}$$

since their number will be decreased only if a whole pill is drawn. On the other hand, for half pills,

$$y_{k+1} - y_k = \frac{x_k}{x_k + y_k} - \frac{y_k}{x_k + y_k} \tag{9.39b}$$

since if a half pill is drawn, the number of halves decrease by one, but increase by one when a whole pill is drawn. Introducing the expected percentages $p_k = x_k/N$ and $q_k = y_k/N$ of whole and half pills remaining, (9.39) becomes

$$\frac{p_{k+1} - p_k}{1/N} = -\frac{p_k}{p_k + q_k} \tag{9.40a}$$

$$\frac{q_{k+1} - q_k}{1/N} = \frac{p_k}{p_k + q_k} - \frac{q_k}{p_k + q_k} \tag{9.40b}$$

so it is reasonable to leap to the model

$$\dot{p} = -\frac{p}{p+q} \tag{9.41a}$$

$$\dot{q} = \frac{p}{p+q} - \frac{q}{p+q} \tag{9.41b}$$

subject to

t	p	q
0	1	0

This system is solved in Exercise 9.29. The lesson to take away from this problem is that *a statistical, indeterminate phenomenon may on average be smoothly determined by a deterministic ODE.*

Problem. Why do minor disturbances bring a congested expressway to a dead stop?

Solution. Suppose that the expressway is in steady state with the kth vehicle at $x_k = x_k(t)$, all drivers following at the distance $d_k = d$ from the car in front, all traveling at the posted speed $v_k = v$. But now a disturbance occurs up the line at $t = 0$. Let us see how this disturbance propagates back up the expressway.

The driver of each following vehicle will tend to accelerate to maintain the same separation distance, to hold $d_k(t) = x_{k-1}(t) - x_k(t)$ constant, i.e., to keep $\dot{x}_{k-1} - \dot{x}_k = 0$. But all drivers need time to react. Thus one might leap to the commonly used *General Motors first model* [Gardner et al.]:

$$\ddot{x}_k(t + \tau) = \alpha[\dot{x}_{k-1}(t) - \dot{x}_k(t)], \tag{9.42}$$

where experimentally on average $\tau = 1.55$ s and $\alpha = 0.37$ s^{-1} [May, p. 168].

But then for future separation distances

$$d_k(t) = x_{k-1}(t) - x_k(t) = \int_0^t [v_{k-1}(s) - v_k(s)]\, ds + d$$

$$= \frac{1}{\alpha} \int_0^t \dot{v}_k(s + \tau)\, ds + d = \frac{v_k(t + \tau)}{\alpha} + d - \frac{v}{\alpha}. \tag{9.43}$$

Separation distances d_k must remain positive to avoid collisions, so

$$d > \frac{v}{\alpha}. \tag{9.44}$$

For our average observed values this means separation $d = 88/0.37 \approx 2.7 \times 88 = 238$ feet at 60 mph (88 ft/s). This 2.7 seconds is consistent with the commonly heard "three-second rule," where the recommended separation distance is the distance traversed in 3 seconds at present speed. Rush hour density is often observed to be 60 vehicles per mile per lane, a spacing of only 88 feet between vehicles. Thus the conservative rule (9.44) would restrict speed to 22 mph. At 30 mph, a delay of $\tau = 1.55$ s for the following driver to respond to braking drops separation to $88 - 44 \times 1.55 = 19.8$ feet, which is a probable collision when vehicle lengths are factored in. It is interesting now to examine observed reaction and braking distances in tests conducted by the U.S. Bureau of Public Roads (Table 9.1).

Table 9.1 [Giordano et al.].

Speed (mph)	Average driver reaction dist. (ft)	Average braking distance (ft)	Average total stopping dist. (ft)
20	22	20	42
30	33	41	74
40	44	72	116
50	55	118	173
60	66	182	248
70	77	266	343
80	88	376	464

The values $\tau = 1.55$ s and $\alpha = 0.37$ s^{-1} in (9.42) and Table 9.1 are based on experiment. At 60 mph Table 9.1 indicates a reaction time of $66/88 = 0.75$ s, half of the delay $\tau = 1.55$ s. The explanation lies in the design of the two experiments (Exercise 9.31).

The mathematical model (9.42) contains a hidden richness of behavior — analysis of only two vehicles is interesting in its own right. Introduce two coordinate systems that move at steady speed v along side two contiguous vehicles:

$$\text{leading vehicle: } u(t) = x_{k-1}(t) - vt - x_{k-1}(0) \tag{9.45a}$$

$$\text{trailing vehicle: } y(t) = x_k(t) - vt - x_{k-1}(0) + d, \tag{9.45b}$$

so that the model (9.42) becomes

$$\ddot{y}(t + \tau) = \alpha[\dot{u}(t) - \dot{y}(t)] \tag{9.46}$$

subject to $y(0) = \dot{y}(0) = 0 = u(0) = \dot{u}(0)$. Integrating (9.46) yields

$$\dot{y}(t + \tau) + \alpha y(t) = \alpha u(t). \tag{9.47}$$

The analysis of this delay equation is best done with the operational (frequency-domain) methods of the Chapter 10, where the model (9.47) becomes

$$Y = \frac{\alpha e^{-\tau s}}{s + \alpha e^{-\tau s}} U. \tag{9.48}$$

We will see that when bad road conditions or a lapse of alertness result in $\tau\alpha > \pi/2$, any perturbation in the velocity of the lead vehicle will precipatate a collision — the two-vehicle system is unstable! By cascading the two-vehicle solution (9.48), we will also see that the farther back in a platoon of vehicles, the larger the excursions in velocity induced by the leading vehicle. All this substantiates the common observation that *delay in a feedback loop is dangerous.* These modeling issues have taken center stage during the present intense interest in intelligent vehicle highway systems (IVHS). We shall return to this problem in later chapters.

See the large collection of differential equation models in the encyclopedic *Mathematical Modeling for Industry and Engineering* by Svobodny. See also the delightful *Introduction to Applied Mathematics* by Strang.

Exercises

9.1 Push a square plate of unit area through a fluid at a moderate but constant normal velocity v. Set the work done against the fluid resistance R equal to the kinetic energy imparted to the volume of fluid in front of the plate to obtain the reasonably accurate relation

$$R = \frac{\rho v^2}{2}$$

where ρ is the density of the fluid.

9.2 Complete the failed furnace problem (9.15).

9.3 A can of pop at 34°F is removed from the refrigerator, placed on a counter, and forgotten. After 2 minutes the can has risen to 35°F. Predict the future evolution of the temperature $T(t)$ of the pop as it warms to the room temperature of 70°F.

9.4 (**Project**) A pipe with cross-sectional area A and circumference C is carrying hot fluid of density ρ and specific heat c at temperature $T(x)$, where x is the distance down the length of the pipe. The hot fluid is moving at constant velocity v. The pipe is uninsulated and is losing heat to ambient. Because of forced convection the surrounding fluid temperature T_a remains constant. Write an ODE for the temperature $T(x)$ at all points down the pipe. Solve and test your result against an actual pipe carrying hot air or fluid.

Answer: $c\rho A v T' = -UC(T - T_a)$, where U is the coefficient of heat transfer to ambient.

9.5 Model the water temperature $T(x)$ at height x within a solar hot water collector. Assume that the collector is a shallow inclined glazed rectangular box well-insulated on the remaining sides, and that the flow is from bottom to top. Try to capture all factors in the model. See [Duffie and Beckman].

9.6 A round object of radius r, mass m, and moment of inertia I is rolling down a vertical wall without slipping — think of a yo-yo. Model its motion.

9.7 A sleeping cat rolls off a window planter and falls toward the pavement far below. Write an ODE for the velocity v of this hapless cat using Newton's rule. Without solving, deduce from Picard's theorem that v increases but never surpasses a limiting *terminal velocity*. What properties of the cat determine this terminal velocity? (According to an apocryphal report by a Manhattan veterinarian there is a "zone of death" between the 2nd and 4th floors — cats that fall from floors outside this zone survive. How can this be?)

9.8 A coasting rowboat experiences subsurface resistance proportional to v^2 but bow-wave surface resistance proportional to v^3. Model the velocity v of this craft. Will it ever come to rest?

9.9 You invest \$1000 in an account earning 5% interest per year compounded continuously. During the year inflation rises steadily from 4% to 6%. Did you make or lose money?

9.10 Establish (9.27) via Euler's formula, $e^{\alpha+i\theta} = e^{\alpha}(\cos\theta + i\sin\theta)$.

9.11 Find the natural frequency of a bobbing cylindrical buoy.

9.12 Show that within the crust of the earth at a distance r from the center, the acceleration a due to gravity is determined only by the material of radius less than r. (The superposition of all accelerations from particles above add to 0.) Thus, assuming the earth to be of more or less uniform density, deduce that $a = -kr$.

9.13 Estimate how long before the system $\ddot{y} + (1/10)\dot{y} + 4y = 0$ with initial values $y(0) = 1, \dot{y}(0) = 0$ has decayed below amplitude 0.01.

Hint: Look at the exponential decay envelope.

Answer: $t \approx 92$ seconds.

9.14 Suppose that at solar noon one person measures the elevation of the sun as 50.3000^{o} while another 40 miles north reads 49.7218^{o}. What is the radius of the earth?

Answer: 3964 miles.

9.15 Using Exercise 9.10, write down the general solution of $(D - 5)^4(D^2 + D + 1)^2 y = 0$ in terms of real-valued functions.

9.16 (**Project**) Does the logistics equation (9.16) model well the population growth in the United States since 1945? If so, what is the carrying capacity? Consult the *Statistical Abstracts of the United States* (an annual publication of the U.S. Bureau of the Census).

9.17 A chain of length L is slipping off a frictionless table. Write an ODE for the portion $x = x(t)$ hanging off the table.

Answer: $\ddot{x} = (g/L)x$.

9.18 A mass m is connected to a spring with constant k by a massless rope draped over a pulley of radius r and moment of inertia I.

Displacement $x = x(t)$ is measured from equilibrium. Write an ODE satisfied by x.

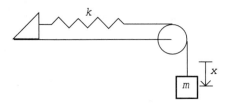

9.19 Model a two-mass planar pendulum by both Newton's and Hamilton's method. Check that your two solutions are consistent.

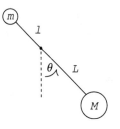

9.20 Model the flow of an incompressible fluid through a tube of varying cross-sectional area A:

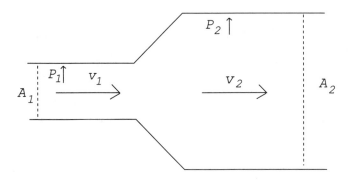

to obtain Bernoulli's relation between pressure P, velocity v, and density ρ:

$$P_1 - P_2 = \frac{\rho}{2} \cdot (v_2^2 - v_1^2).$$

Outline: Constant throughput yields $A_1 v_1 = A_2 v_2$. Now equate the sum of kinetic energy and work done within the two portions of the tube.

Sketch why an aircraft flies. Why does a sail take on the airfoil shape?

9.21 Find a three-dimensional system modeling the trajectory of a shoulder-fired stinger surface-to-air missile as it rises to destroy an unsuspecting airbus.

Hint: First do the case where gravity and the winds aloft are neglected. Assume constant missile speed and adapt (9.34). Then attempt to factor in gravity, winds, and the decreasing mass of the missile as it burns off fuel.

9.22 Set $\psi = (q,p)^T$ and define the matrices

$$H = \begin{pmatrix} k & 0 \\ 0 & 1/m \end{pmatrix} \text{ and } J = \begin{pmatrix} 0 & -1 \\ 1 & 0 \end{pmatrix}$$

to transform the harmonic motion phase space model (9.32) with $c = 0$ into a Schrödinger-like model,

$$J\dot\psi = H\psi$$

where $J^2 = -I$. Even more suggestively, find matrices T and V so that $H = T + V$, $\langle T\psi, \psi \rangle /2 = p^2/2m$ (kinetic energy), and $\langle V\psi, \psi \rangle /2 = kq^2/2$ (potential energy).

9.23 Argue (9.33a).

9.24 By introducing the states $q_i = x_i$ and $p_i = m_i \dot x_i$, write the second-order spring–mass n-system (9.33) as a first-order $2n$-system.

9.25* By choosing $M = \text{diag}(m_i)$ and $K = \text{tridiag}(-k_{i-1}, k_{i-1} + k_i, -k_i)$ write the multiple mass–spring system (9.33a) in the matrix form (9.33b):

$$M\ddot x + Kx = 0.$$

Prove K *positive definite;* i.e., $\langle Kx, x \rangle = x^T Kx > 0$ for $x \neq 0$.

9.26 Sketch together the temperatures $t = t(x)$ and $T = T(x)$ of Figure 9.9. Why must the two curves never cross? Give two arguments — one physical, one mathematical. Determine Fourier's four famous constants b, B, d, D in (9.38).

9.27 (**Project**) Why are heat exchangers almost always counterflow? Compare the efficiency of a co-current versus a counterflow heat exchanger. Use actual heat transfer coefficients for copper tubing of standard thickness. Look up heat transfer coefficients at various fluid speeds. Discuss the improvement due to flow turbulence versus the cost of increased pumping power.

9.28 How must the model of the counterflow heat exchanger (9.37) be modified if the outer tube is uninsulated and losing heat to ambient?

9.29 Solve the heartworm system (9.41).

Hint: $dq/dp = \dot{q}/\dot{p} = -1 + q/p$, so $q = -p \ln p$.

Answer: $p - p \ln \sqrt{p} = 1 - t/2$.

9.30 (**Project**) Take data on expressway traffic flow. What is the distribution of time *headway* between cars at various times of the day? What is their separation versus velocity? How is flow rate related to density? When is this road most dangerous? A major part of this project is a clever design of how you will obtain your data. Much can be gleaned with a stopwatch and premeasured portions of road.

9.31 How can the delay time of $\tau = 1.55$ s seen in experimental verification of (9.42) by General Motors be reconciled with the reaction times in Table 9.1 taken by the U.S. Bureau of Public Roads? The former data was taken in traffic, the latter on a test range.

9.32 Using the data from Table 9.1 and assuming that braking produces a linear decrease in speed, formulate your own rule of thumb for a safe following distance.

9.33*Possibly the most famous ODE system is the *Lotka–Volterra predator–prey*:

$$\dot{x} = ax - bxy$$
$$\dot{y} = -cy + dxy,$$

where y is the number of predators, x the number of prey, $a, b, c, d > 0$. Think of a pond of pike and bluegill. Prove that the first-quadrant solution trajectories $(x(t), y(t))$ are closed loops by proposing an "energy" function that is an *integral*, i.e., constant along trajectories.

Hint: Try $E = x + y - \ln xy$ when $a = b = c = d = 1$.

9.34 Using the given solutions to damped harmonic motion (9.28), graph various phase space trajectories $(q(t), p(t))$ of (9.32) for various damping coefficients ζ.

9.35 Why do flags flap rather than stream in moderate breezes?

9.36 Recall that the general solution y to a linear but nonhomogeneous ODE $Lu = f$ is the sum $y = y_p + y_h$ of a particular solution y_p of the nonhomogeneous equation plus the general solution y_h of the homogeneous $Lu = 0$. Find the general solution of

$$y'' + 3y' + 2y = 2.$$

9.37 Find a particular solution y_p of

$$y'' + y' + 3y = 2 - x + 5x^2$$

by trying solutions of the form

$$y = A + Bx + Cx^2.$$

9.38 Find the general solution of

$$\ddot{y} + \omega^2 y = \cos \alpha t$$

first for $\omega \neq \alpha$, then for $\omega = \alpha$.

9.39 Two armies are battling to the death. Each army is being decimated at a rate proportional to the size of the other while being replenished at a constant rate. What is the outcome — stalemate or annihilation? What if each army is being replenished at a rate proportional to its present size?

9.40 Solve the *Euler type* equation

$$x^2 y'' - 2xy' + 2y = 0$$

by attempting solutions of the form $y = x^r$.

9.41 A linear system of the form

$$\dot{x} = Ax$$

subject to $x(0) = x_0$, where x is a $n \times 1$ time-varying vector and A a $n \times n$ matrix of constants, is solved by applying the *transition matrix*

$$T(t) = e^{At} = I + \frac{t}{1!}A + \frac{t^2}{2!}A^2 + \frac{t^3}{3!}A^3 \cdots$$

to the initial position x_0 to obtain the solution trajectory

$$x(t) = T(t)x_0.$$

Using this approach solve the system

$$\dot{x} = \begin{pmatrix} 1 & 1 \\ 0 & 1 \end{pmatrix} x$$

subject to

t	x_1	x_2
0	1	−1

9.42 Solve $y'' - xy' - y = 0$ by trying solutions of the form

$$y = \sum_{k=0}^{\infty} c_k x^k.$$

Answer:

$$y = c_0\left(1 + \frac{x^2}{2} + \frac{x^4}{2 \cdot 4} + \cdots\right)$$

$$+ c_1\left(x + \frac{x^3}{3} + \frac{x^5}{3 \cdot 5} + \cdots\right).$$

Chapter 10

Frequency-Domain Methods

Enormous computational power and insight is achieved with the operational methods of Oliver Heaviside.

We first review the Laplace transform method for solving ODEs, only to see that this transform approach was merely an early attempt to employ generalized functions. The efficacy of the frequency domain becomes clear when plants are cascaded. You will learn how to model electronic circuits via Heaviside's method. Several criteria for deciding the stability of plants are proved and practiced. We return to Bode plots and the design of lowpass, bandpass, and highpass filters. Feedback is illustrated by exploiting the all-purpose op amp. Stabilizing plants with feedback is graphically demonstrated with the root-locus method, followed by a short introduction to Nyquist analysis. The chapter ends with several control problems: how feedback can improve the performance of a servomechanism by restricting overshoot and improving response time, and in the analysis of an automobile cruise control system. We conclude with a stability analysis of the ill-fated submarine C.S.S. *Hunley*.

§10.1 The Frequency Domain

Oliver Heaviside's powerful computational methods were at first justified as Laplace transform techniques [Nahin]. *Ordinary signals* in the *time domain*, locally integrable functions $f(t)$ of time t that are quiescent (vanishing) before $t = 0$, are transformed via the Laplace transform

$$F(s) = \int_0^\infty f(t)e^{-st}\,dt \qquad (10.1)$$

149

to their *frequency-domain* alter ego $F(s)$, manipulated, then transformed back to the time domain.

Example. The ODE $\ddot{y}+3\dot{y}+2y = 1$ subject to zero initial conditions transforms to

$$s^2Y + 3sY + 2Y = \frac{1}{s}$$

so

$$Y = \frac{1}{s(s+1)(s+2)} = \frac{A}{s} + \frac{B}{s+1} + \frac{C}{s+2}. \qquad (10.2)$$

Multiplying (10.2) through by each factor $s - \alpha$ and then setting $s = \alpha$ in turn gives $A = 1/2$, $B = -1$, $C = 1/2$, i.e.,

$$Y = \frac{1/2}{s} - \frac{1}{s+1} + \frac{1/2}{s+2},$$

whereupon transforming each term back to the time domain gives our solution,

$$y = (1/2) - e^{-t} + (1/2)e^{-2t}.$$

See Tables 10.1 and 10.2 for rules and formulas for the Heaviside methods. See [Råde and Westergren] for more complete tables.

Table 10.1. Rules

Time domain	Frequency domain
$f(t)$	$F(s)$
$af(t) + bg(t)$	$aF(s) + bG(s)$
$\dot{f}(t)$	$sF(s) - f(0)$
$f^{(n)}(t)$	$s^n F(s) - s^{n-1}f(0) - \cdots - f^{(n-1)}(0)$
$t^n f(t)$	$(-1)^n F^{(n)}(s)$
$f(t-a)$	$e^{-as}F(s)$
$e^{-at}f(t)$	$F(s+a)$
$\int_0^t f(\tau)\, d\tau$	$F(s)/s$
$f(t) * g(t)$	$F(s)G(s)$
$f(at)$	$F(s/a)/a$

Table 10.2. Formulas

Time domain	Frequency domain
1	$1/s$
t^n	$n!/s^{n+1}$
e^{-at}	$1/(s+a)$
$\cos at$	$s/(s^2+a^2)$
$\sin at$	$a/(s^2+a^2)$
$\cosh at$	$s/(s^2-a^2)$
$\sinh at$	$a/(s^2-a^2)$
$J_0(at)$	$1/\sqrt{s^2+a^2}$
$\delta(t)$	1

Problem. Solve the system

$$\dot{x} = -2x - 9y \qquad (10.3\text{a})$$

$$\dot{y} = x - 2y \qquad (10.3\text{b})$$

subject to $x(0) = -1$, $y(0) = 1$.

Solution. Transforming (10.3) to the frequency domain, we obtain

$$sX + 1 = -2X - 9Y$$

$$sY - 1 = X - 2Y,$$

i.e.,

$$(s+2)X + 9Y = -1$$

$$-X + (s+2)Y = 1,$$

so by Cramer's rule,

$$X = \frac{\begin{vmatrix} -1 & 9 \\ 1 & s+2 \end{vmatrix}}{\begin{vmatrix} s+2 & 9 \\ -1 & s+2 \end{vmatrix}} = -\frac{s+2+9}{(s+2)^2+3^2}$$

$$= -\frac{s+2}{(s+2)^2+3^2} - \frac{9}{(s+2)^2+3^2}.$$

Consulting Tables 10.1 and 10.2, we see that

$$x = -e^{-2t}\cos 3t - 3e^{-2t}\sin 3t. \qquad (10.4\text{a})$$

Similarly (Exercise 10.3),

$$y = e^{-2t} \cos 3t - (1/3)e^{-2t} \sin 3t. \tag{10.4b}$$

Systems can also be solved operationally in matrix form: the system $\dot{z} = Az$ subject to $z(0) = z_0$, where A is an $n \times n$ matrix of constants, when taken to the frequency domain becomes $sZ - z_0 = AZ$, so the solution in the frequency domain is

$$Z = (sI - A)^{-1} z_0,$$

where the matrix $R(s) = (sI - A)^{-1}$ is called the *resolvent* of the system. Because of the fundamental identity

$$M \cdot \mathrm{adj}(M) = \det(M)I$$

for any square matrix M [Brown, p. 16], each entry of the resolvent $R(s)$ is a rational function in s with denominator $\det(sI - A)$ and numerator of degree at most $n - 1$. Hence the resolvent has a matrix partial fraction expansion

$$(sI - A)^{-1} = \sum_{k=1}^{r} \sum_{j=1}^{m_k} \frac{A_{jk}}{(s - \beta_k)^j}, \tag{10.5}$$

where $\beta_1, \dots, \beta_r$ are the distinct roots of the characteristic polynomial

$$\det(sI - A) = \prod_{k=1}^{r} (s - \beta_k)^{m_k}.$$

Taking (10.5) back to the time domain and applying it to the vector z_0 yields the solution.

Example. The 2×2 system

$$\dot{z} = \begin{pmatrix} 2 & 3 \\ 2 & 1 \end{pmatrix} z$$

subject to $z(0) = z_0 = (1, 1)^T$ has frequency-domain solution

$$Z = \begin{pmatrix} s - 2 & -3 \\ -2 & s - 1 \end{pmatrix}^{-1} z_0.$$

But the resolvent

$$R(s) = \begin{pmatrix} s-2 & -3 \\ -2 & s-1 \end{pmatrix}^{-1} = \frac{1}{(s-4)(s+1)} \begin{pmatrix} s-1 & 3 \\ 2 & s-2 \end{pmatrix}$$

$$= \frac{B}{s-4} + \frac{C}{s+1} = \text{(Exercise 10.4)}$$

$$= \frac{1}{s-4} \begin{pmatrix} 3/5 & 3/5 \\ 2/5 & 2/5 \end{pmatrix} + \frac{1}{s+1} \begin{pmatrix} 2/5 & -3/5 \\ -2/5 & 3/5 \end{pmatrix}. \qquad (10.6)$$

Thus returning to the time domain,

$$z = e^{4t} \begin{pmatrix} 3/5 & 3/5 \\ 2/5 & 2/5 \end{pmatrix} z_0 + e^{-t} \begin{pmatrix} 2/5 & -3/5 \\ -2/5 & 3/5 \end{pmatrix} z_0 = \frac{1}{5} \begin{vmatrix} 6e^{4t} - e^{-t} \\ 4e^{4t} + e^{-t} \end{vmatrix}.$$

Note that this approach (in contrast to the last) puts the solution in hand for any initial condition z_0.

*§10.2 Generalized Signals

In the last several decades it has become clear that the Laplace transform explanation of Heaviside's methods is not the correct explanation. As laid out in 1959 by Jan Mikusiński in his important book *Operational Calculus*, the frequency domain is merely the time domain enlarged to include *generalized* signals. It is exactly analogous to the enlargement of the domain of integers **Z** to its field of quotients **Q** of rational numbers.

Consider the domain of all *causal ordinary signals* $f(t)$, i.e., all real- or complex-valued functions $f(t)$ of time t with $f(t) = 0$ when $t < 0$ and

$$\int_0^t |f(\tau)| \, d\tau < \infty$$

for all t. These ordinary signals form a commutative ring under pointwise addition and *convolution* $*$, where $h = f * g$ means

$$h(t) = \int_0^t f(\tau)g(t - \tau) \, d\tau. \qquad (10.7)$$

By a famous result of Titchmarsh [Erdelyi], this ring of ordinary signals is free of zero divisors, i.e., $f * g = 0 \Rightarrow f = 0$ or $g = 0$. Thus we may successfully define a field of *convolution quotients* consisting of all formal quotients $f/{}^*g$ of ordinary signals with $g \neq 0$, where equality is defined by

$$f_1/{}^*g_1 = f_2/{}^*g_2 \iff f_1 * g_2 = f_2 * g_1.$$

These new objects are called *generalized signals.* The ordinary signals are found to be naturally embedded as a subring of the generalized signals by the identification

$$f(t) \longrightarrow \{f(t)\} = (f * g)/{}^*g.$$

For ordinary signals f and g, in this notation

$$\{f(t)\} \cdot \{g(t)\} = \{f(t) * g(t)\}.$$

Example. The generalized signal $g/{}^*g$ is denoted by δ or 1 and is called the *Dirac delta function.* The Dirac delta acts as the multiplicative identity of the field of generalized signals. This is clearly a new object — there is no ordinary signal $d(t)$ with $d(t) * f(t) = f(t)$ for all, or even some, nonzero ordinary f (Exercise 10.6).

Example. The familiar *Heaviside unit step* is the ordinary signal

$$u_0(t) = \begin{cases} 1 & \text{if } t \geq 0 \\ 0 & \text{otherwise.} \end{cases}$$

Let h denote the embedded image of the unit step among the generalized signals

$$h = \{u_0(t)\}.$$

Each generalized signal acts as an operator on all generalized signals by left multiplication. Multiplication by the unit step h is called *generalized integration* since

$$h \cdot \{f(t)\} = \{u_0(t) * f(t)\} = \Big\{ \int_0^t f(\tau) \, d\tau \Big\}.$$

Example. Introduce the generalized signal $s = 1/h$. Multiplication by this operator is called *generalized differentiation,* since it inverts generalized integration. (Heaviside used the letter p for generalized differentiation; the symbol s is inherited from the Laplace transform.)

Many transcendental ordinary functions can be rewritten very simply in terms of generalized differentiation s.

Example. Since

$$h \cdot \{e^{-at}\} = \{\int_0^t e^{-a\tau} \, d\tau\} = \{\frac{1 - e^{-at}}{a}\}$$

$$= \{\frac{1}{a}\} - \{\frac{e^{-at}}{a}\} = \frac{h}{a} - \frac{\{e^{-at}\}}{a},$$

multiplying through by s and a yields

$$a\{e^{-at}\} = 1 - s\{e^{-at}\},$$

so

$$\{e^{-at}\} = \frac{1}{s+a}. \tag{10.8}$$

All the familiar transforms of Table 10.2 are obtained in like manner, *free of convergence questions*. Convergence of the Laplace transform is irrelevant. Even $f(t) = e^{t^2}$, whose Laplace transform converges nowhere, has a 'transform,' viz., $\{e^{t^2}\}$.

Example. Delay requires some care. The identification $\{f(t - a)\} = e^{-as} F(s)$ presupposes a notion of convergence on the generalized functions — see [Erdelyi] or [Mikusiński] for the details. But once convergence is defined, it is easy to deduce the identification $\{tf(t)\} = -F'(s)$, and from this the operation of delay: Let $u_a(t) = u_0(t - a)$, the unit step delayed by a seconds, and let $Y(s) = \{u_a(t)\}$. Then since

$$tu_a(t) = au_a(t) + \int_0^t u_a(\tau) \, d\tau,$$

$$-Y(s)' = aY(s) + Y(s)/s,$$

so

$$(sY(s))' + a(sY(s)) = 0,$$

giving (Exercise 10.8)

$$\{u_a(t)\} = \frac{e^{-as}}{s}$$

and hence (Exercise 10.8)

$$\{f(t-a)\} = e^{-as}F(s). \tag{10.9}$$

A *plant* is a linear time-invariant causal mapping of generalized signals to generalized signals. Once convergence is defined for the field of generalized signals, one writes any signal $u(t)$ as a superposition of delays, viz.,

$$U = \int_0^\infty u(\tau)\delta(t-\tau)\,d\tau = \int_0^\infty u(\tau)e^{-s\tau}\,d\tau.$$

Let p be the (causal) *impulse response*, the response of the plant to the input δ. Then by linearity we may extend the action of the plant over the integral. By time invariance, $\delta(t-\tau)$ is mapped to $p(t-\tau)$. Thus the response of the plant to u is

$$Y = \int_0^\infty u(\tau)p(t-\tau)d\tau = p*u. \tag{10.10a}$$

In short, *a plant is given by multiplication (convolution) by a fixed generalized signal called its impulse response.* In older notation,

$$Y(s) = P(s)U(s), \tag{10.10b}$$

where u is the *input* and y the *output*. From this, more modern point of view, *the transfer function and the impulse response coincide.*

Example. Consider the plant defined by the ODE $\dot{y}+ay = u$, where it is understood that the input $u(t)$ and output $y(t)$ are causal, i.e., vanishing for $t < 0$. Multiplying through by the integrating factor e^{at}, we see that $(e^{at}y)' = e^{at}u$, so

$$y(t) = e^{-at}\int_0^t e^{a\tau}u(\tau)\,d\tau + y(0)e^{-at} = \int_0^t e^{-a(t-\tau)}u(\tau)\,d\tau + y(0)e^{-at}.$$

Insisting on a continuous solution for continuous inputs $u(t)$ forces $y(0) = 0$. Thus the plant is convolution by the impulse response e^{-at},

$$y(t) = e^{-at}*u(t).$$

Extending to all generalized inputs U, the plant is given by

$$Y = \frac{U}{s+a},$$

i.e., multiplication by $1/(s+a)$.

Example. The ideal linear amplifier of gain k given by $y = ku$ is convolution with the generalized signal $k\delta$.

§10.3 Plants in Cascade

The frequency domain is especially useful when cascading plants.

Example. Consider a plant composed of an inverting linear amplifier of gain -5, followed by an integrator, followed in turn by convolution with e^{-2t}, and finally by a 3-second delay (Figure 10.1). The time-domain version of this compound plant is nasty (Exercise 10.10), but in the frequency domain the transfer function is simply the product of the cascaded plants:

$$\frac{Y}{U} = -5 \cdot \frac{1}{s} \cdot \frac{1}{s+2} \cdot e^{-3s} = \frac{-5e^{-3s}}{s(s+2)}. \qquad (10.11)$$

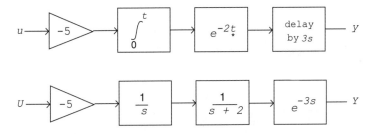

Figure 10.1. Plants in cascade.

Problem. A person is shouting a signal $u(t)$ across a canyon (Figure 10.2). The opposite canyon face is 1 second away at the speed of sound (1087 ft/s). The canyon walls reflect with efficiency 10%. A microphone directly in front of this person is recording the signal $y(t)$. Find y.

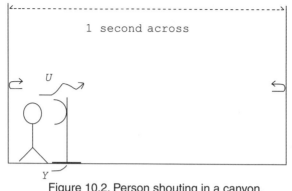

Figure 10.2. Person shouting in a canyon.

Solution. We assume the typical cardioid spatial response pattern for the microphone — it does not hear from the back. Let us abstract this system as a block diagram, as shown in Figure 10.3. Thus computing Y in two ways and equating,

$$Y = U + e^{-s} \cdot 0.1e^{-s} \cdot 0.1Y,$$

so

$$\frac{Y}{U} = \frac{1}{1 - 0.01e^{-2s}} = 1 + 10^{-2}e^{-2s} + 10^{-4}e^{-4s} + 10^{-6}e^{-6s} + \cdots,$$

thus verifying the intuitive guess that a short shout yields a train of echos, each delayed by 2 seconds and decreased in amplitude by $1/100$ from the previous echo.

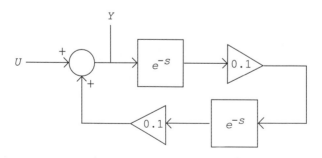

Figure 10.3. Canyon system of Figure 10.2.

Example. Let us cascade two plants, the first a lowpass filter given by $\dot{y} + y = u$ and the second a highpass filter given by $\dot{z} + z = \dot{y}$

(Figure 10.4). In the frequency domain the solution is simple:

$$sY + Y = U \quad \text{and} \quad sZ + Z = sY.$$

Thus

$$Z = \frac{sY}{1+s} = \frac{sU}{(s+1)^2}$$

with transfer function

$$\frac{Z}{U} = \frac{s}{(s+1)^2} = \frac{1}{s+1} - \frac{1}{(s+1)^2}.$$

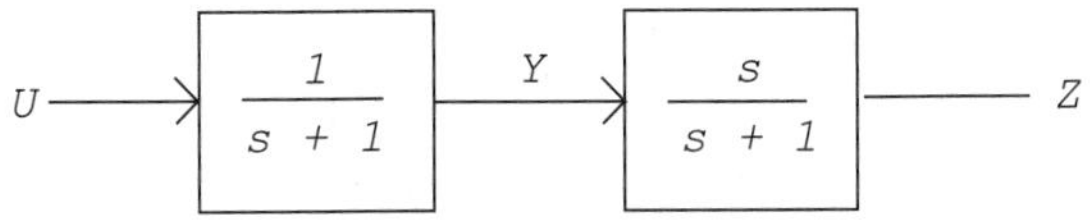

Figure 10.4. Two cascaded filters.

So the *step response*, the response to the Heaviside unit step $U = 1/s$, is

$$Z = \frac{1}{s} \frac{s}{(s+1)^2} = \frac{1}{(s+1)^2},$$

i.e.,

$$y(t) = te^{-t}. \tag{10.12}$$

§10.4 Surge Impedance

The observed relation between the voltage $v(t)$ *across* and the current $i(t)$ flowing *through* a resistor, capacitor, and inductor is shown in Figure 10.5. The ratio $Z = V/I$ is called the *surge impedance* of a device. Surge impedances in series and parallel satisfy the famous rules illustrated in Figure 10.6 (Exercise 10.14).

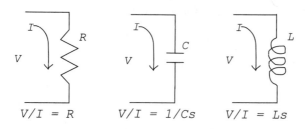

Figure 10.5. Relation of voltage to current
among the three basic components.

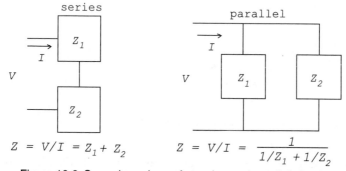

Figure 10.6. Surge impedance for series and parallel circuits.

Problem. What is the transfer function of the (notch) filter shown in Figure 10.7? (The convention is that the output voltage y is not loaded down — there is no current flowing out that port — and initially, no energy is stored in any component.)

Solution. We rewrite the circuit of Figure 10.7 as shown in Figure 10.8, where $Z_1 = R$ and $Z_2 = Ls + 1/Cs$. But in general, for Figure 10.8,

$$\frac{Y}{U} = \frac{Z_2}{Z_1 + Z_2} \qquad (10.13)$$

(Exercise 10.16). Thus our notch has transfer function

$$\frac{Y}{U} = \frac{Ls + 1/Cs}{R + Ls + 1/Cs} = \frac{LCs^2 + 1}{LCs^2 + RCs + 1}$$

and so has the time-domain model

$$LC\ddot{y} + RC\dot{y} + y = LC\ddot{u} + u.$$

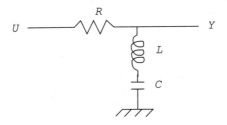

Figure 10.7. Series resonant notch filter.

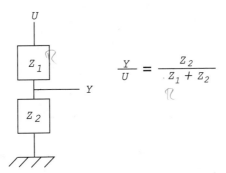

$$\frac{Y}{U} = \frac{Z_2}{Z_1 + Z_2}$$

Figure 10.8. Notch filter of Figure 10.7 in block form.

§10.5 Stability

A plant is *stable* if bounded ordinary inputs yield bounded ordinary outputs.

Stability is much more difficult to characterize for continuous time than for discrete time, where every frequency-domain object comes from an ordinary signal. One successful continuous-time characterization is found in [MacCluer, 1988]: *The stable plants are those given by convolution with the generalized derivative of an ordinary signal of bounded variation.* But fortunately, the most useful case is quite easy to understand and apply.

THEOREM A. A *proper rational* plant

$$\frac{Y}{U} = P(s) = \frac{b_0 s^m + b_1 s^{m-1} + \cdots + b_m}{a_0 s^n + a_1 s^{n-1} + \cdots + a_n} \qquad (10.14)$$

(in lowest terms) with $m \leq n$ and $a_0 \neq 0$ is stable if and only if all its poles (the zeros of the denominator) lie in the open left-hand complex plane Re $s < 0$.

PROOF. Expand $P(s)$ in partial fractions:

$$P(s) = \kappa + \sum_{k=1}^{r} \sum_{j=1}^{m_k} \frac{A_{jk}}{(s - \beta_k)^j}. \qquad (10.15)$$

It is enough then (details omitted) to show the result for

$$P(s) = \frac{1}{(s - \beta)^k},$$

i.e., for convolution by $p(t) = t^{k-1} e^{\beta t}/(k-1)!$. But (Exercise 10.18) $y(t) = t^{k-1} e^{\beta t} * u(t)$ is bounded for all bounded $u(t)$ if and only if Re $\beta < 0$.

Example. The second-order transfer function

$$P(s) = \frac{3s^2 + 2s + 1}{s^2 + 7s + 4}$$

is stable since the coefficients of the denominator are all positive, so (by the quadratic formula) its zeros lie strictly to the left of the imaginary axis (Exercise 10.20).

Observation A. Stable plants with proper rational transfer function (10.15) have an impulse response that is the sum of a linear amplifier (of gain possibly 0) and damped terms of the form $t^k e^{\alpha t}$ with Re $\alpha < 0$. *The impulse response dies away.*

Observation B. In practice, stable proper rational plants $P(s)$ have real coefficients whereupon the terms of (10.15) occur in conjugate pairs that combine into damped sinusoids $t^k e^{-at} \sin(\omega t + \theta)$ with $a > 0$.

Observation C. If the input $u(t)$ to a stable proper real rational plant is a linear combination of sinusoids or more generally of signals $t^k e^{at} \sin(\omega t + \theta)$, the output $y(t)$ is also a linear combination of such signals that can be partitioned into two cohorts:

$$y(t) = \sum \text{damped system modes} + \sum \text{input modes},$$

where the *system modes* are terms arising from the poles of the plant $P(s)$ and where the *input modes* come from the poles of the input $U(s)$. Merely expand $Y(s) = P(s)U(s)$ in partial fractions. Poles common to both are placed in one of the two cohorts.

Plants with an ordinary impulse response enjoy a stability criterion exactly analogous to the one for discrete-time systems (Result B, §3.4).

THEOREM B. Convolution by the ordinary signal $p(t)$ is stable if and only if $p(t)$ is absolutely summable, i.e.,

$$\int_0^\infty |p(t)|\, dt < \infty. \tag{10.16}$$

*PROOF. Suppose $p(t)$ is absolutely summable. Then for any bounded ordinary signal $u(t)$,

$$|y(t)| = |p(t) * u(t)| = |\int_0^t p(\tau)u(t-\tau)\, d\tau| \le \int_0^t |p(\tau)| \cdot |u(t-\tau)|\, d\tau$$

$$\le \|u\|_\infty \cdot \int_0^t |p(\tau)|\, d\tau \le \|u\|_\infty \cdot \int_0^\infty |p(\tau)|\, d\tau,$$

where $\|u\|_\infty = \sup |u(t)|$, $0 \le t < \infty$. Thus bounded inputs yield bounded outputs.

Conversely, suppose $y(t) = p(t) * u(t)$ is stable and hence a linear mapping of $X = L^\infty(0, \infty)$ to itself. Suppose $u_n \to 0$ and $y_n = p * u_n \to y_0$ in X. Then by the first estimate above, $y_n(t) \to 0$ pointwise, so $y_0 = 0$. Thus by the closed graph theorem [Riesz and Sz-Nagy, p. 307], this convolution map is a *bounded* operator on $L^\infty(0, \infty)$, i.e.,

$$|p(t) * u(t)| = |\int_0^t p(\tau)u(t-\tau)\, d\tau| \le \|u\|_\infty \cdot B \tag{10.17}$$

for all u for some bound B belonging to p. But then consider the special bounded inputs of the form $u_\alpha(t) = \operatorname{sgn} p(\alpha - t)$. By (10.17),

$$\left| \int_0^t p(\tau) u_t(t-\tau)\, d\tau \right| = \int_0^t p(\tau)\operatorname{sgn} p(\tau)\, d\tau = \int_0^t |p(\tau)|\, d\tau \le B$$

for all t, giving (10.16).

Example. Convolution by $p(t) = e^{-t^2}$ is stable since $p(t)$ is certainly absolutely summable. Theorem A is of no help since this $p(t)$ is not a rational function in s (Exercise 10.21).

§10.6 Filters

In common parlance the word *filter* refers to (linear, causal, time–invariant) plants used in processing information rather than modeling mechanisms. The plant modeling a backhoe bucket angle in response to the hydraulic lever position would not be called a filter.

In direct analogy with discrete time, filter frequency response is given by evaluating the transfer function along the imaginary axis.

THEOREM C. Suppose that $H(s)$ is the transfer function of the stable filter given by convolution with the (absolutely summable) real ordinary signal $h(t)$. Then after transients die away, the response of this filter to the sinusoid $u(t) = \sin \omega t$ is a sinusoid $y(t) = r \sin(\omega t + \phi)$ of the same frequency but with *gain*

$$r(\omega) = |H(\omega i)| \tag{10.18a}$$

and *phase lead*

$$\phi(\omega) = \arg(H(\omega i)). \tag{10.18b}$$

PROOF. The response to the complex exponential $e^{i\omega t}$ is

$$h(t) * e^{i\omega t} = \int_0^t h(\tau)e^{i\omega(t-\tau)}d\tau = e^{i\omega t}\int_0^t h(\tau)e^{-i\omega\tau}d\tau$$

$$= e^{i\omega t}\left[\int_0^\infty h(\tau)e^{-i\omega\tau}\,d\tau - \int_t^\infty h(\tau)e^{-i\omega\tau}\,d\tau\right]$$

$$= e^{i\omega t}H(\omega i) + o(1),$$

where $o(1)$ denotes a function with limit 0 as $t \to \infty$.

Fix ω and set $H(\omega i) = re^{i\phi}$. Then since $h(t)$ is real,

$$h(t) * \sin \omega t = h(t) * \frac{e^{i\omega t} - e^{-i\omega t}}{2i}$$

$$= \frac{e^{i\omega t}H(\omega i) - e^{-i\omega t}H(-\omega i)}{2i} + o(1) = \frac{re^{i\omega t + i\phi} - re^{-i\omega t - i\phi}}{2i} + o(1)$$

$$= r\sin(\omega t + \phi) + o(1).$$

Theorem C reveals an approach for judging the periodic steady state performance of a filter via its *Bode plots*, the graphs of gain $r = r(\omega)$ and phase $\phi = \phi(\omega)$ against frequency ω.

Example. Let us examine the performance of the notch filter of Figure 10.7 for component values $R = C = L = 1$. The transfer function is

$$H(s) = \frac{Y(s)}{U(s)} = \frac{s^2 + 1}{s^2 + s + 1}. \tag{10.19}$$

Evaluating $H(s)$ at $s = \omega i$ yields $H(\omega i) = re^{i\phi}$, which when graphed against frequency $f = \omega/2\pi$ yields the gain and phase plots of Figure 10.9.

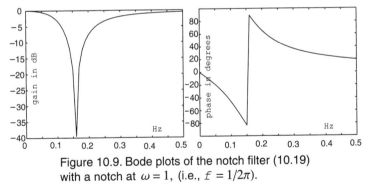

Figure 10.9. Bode plots of the notch filter (10.19) with a notch at $\omega = 1$, (i.e., $f = 1/2\pi$).

These plots were obtained by Routine 10.1:

```
% Routine 10.1  Bode plots
fmin = 0;            % min.  freq.  (in hertz) to view
fmax = 0.5;          % max.  freq.  (in hertz) to view
df=0.01;             % step size
f=fmin:df:fmax;      % frequency vector
```

```
w = 2*pi*f;          % radian freq.  vector
s = w*i;             %
N= s.*s + 1;         % numerator of transfer fn
D = s.*s + s + 1;    % denominator of transfer fn
H = N./D;            % transfer fn
r = abs(H);          % gain at freq.  w
phi = angle(H);      % phase lead
plot(f,r)            % gain vs freq (in hertz)
% replace with      plot(f,phi) for phase plot
% replace with      plot(f,20*log10(r)) for decibels
% replace with      plot(w,r) for gain vs.  radian freq.  w
```

When designing a stable proper rational filter,

$$H(s) = \frac{b_0 s^m + b_1 s^{m-1} + \cdots + b_m}{a_0 s^n + a_1 s^{n-1} + \cdots + a_n},$$

follow these rules:

1. To reduce the gain of the filter at a frequency ω_0, place two conjugate zeros $-\sigma \pm i\omega_0$ either on or close to the imaginary axis, i.e., take $|\sigma|$ small.

2. To enhance the response at a frequency ω_0 place two stable conjugate poles $-\sigma \pm i\omega_0$ close but to the left of the imaginary axis, with $\sigma > 0$ and small.

Problem. Design a bandpass filter to pass the frequencies between 10 and 15 Hz but to suppress all others.

Solution. Let us cascade two second-order filters of the form

$$F(s) = \frac{s/Q}{s^2 + s/Q + 1}. \tag{10.20}$$

According to Exercise 10.29, this filter $F(s)$ of (10.20) peaks with gain 1 at radian frequency $\omega = 1$ and falls off on either side of $\omega = 1$ more rapidly with larger Q (Figure 10.10). Moreover, such filters enjoy minimal phase distortion with larger Q — see Exercise 10.30.

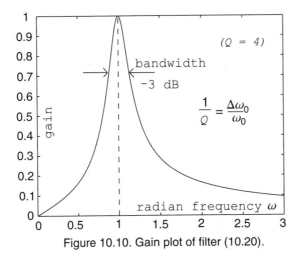

Figure 10.10. Gain plot of filter (10.20).

We may move this filter (10.20) from the normalized radian frequency $\omega = 1$ by the scaling $s \to s/\omega_0$ to obtain a filter centered at $\omega = \omega_0$:

$$F(s) = \frac{\omega_0 s/Q}{s^2 + \omega_0 s/Q + \omega_0^2}. \tag{10.21}$$

Let us cascade two filters of this type, one centered at 10 Hz ($\omega_0 = 20\pi$), the other at 15 Hz ($\omega_0 = 30\pi$) with a $Q = 4$:

$$L(s) = \frac{20\pi s/4}{s^2 + 20\pi s/4 + 400\pi^2} \tag{10.22a}$$

and

$$R(s) = \frac{30\pi s/4}{s^2 + 30\pi s/4 + 900\pi^2}. \tag{10.22b}$$

Finally, cascade these two to form

$$H(s) = 3.3L(s)R(s). \tag{10.23}$$

A simple modification (Exercise 10.31) of Routine 10.1 yields the Bode gain plot shown in Figure 10.11 for our bandpass filter (10.23).

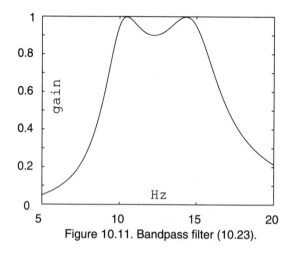

Figure 10.11. Bandpass filter (10.23).

Question. How does one actually build the filter sections

$$F(s) = \frac{\omega_0 s / Q}{s^2 + \omega_0 s / Q + \omega_0^2}$$

used in the bandpass filter above?

Answer 1. Manipulate the transfer function $F(s)$ until it is composed of series and parallel blocks of recognizable surge impedances:

$$F(s) = \frac{\omega_0 / Q}{s + \omega_0 / Q + \omega_0^2 / s}.$$

Referring to Figure 10.8, we see that $Z_2 = \omega_0/Q = R$ and $Z_1 = s + \omega_0^2/s = Ls + 1/Cs$, which is the passive circuit of Figure 10.12.

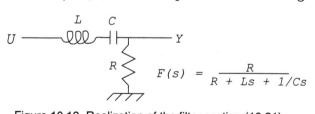

Figure 10.12. Realization of the filter section (10.21).

Answer 2. Nowadays these filters are often "computed" using operational amplifiers as in Exercise 10.38, (see the *Active Filter Cookbook* by Lancaster).

§10.7 Feedback and Root Locus

Feedback is used to steer a process in a more favorable direction.

Example. One of the standard building blocks of modern electronics is the *operational amplifier,* an extremely high gain $\beta \approx 10^8$ linear amplifier with differential inputs (Figure 10.13).

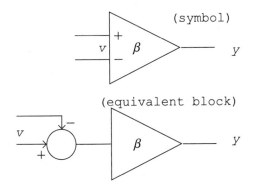

Figure 10.13. Op amp: a high-gain differential input linear amplifier.

The inputs are essentially of infinite impedance, sinking no current. A voltage potential v appearing across the differential inputs is amplified to an output voltage $y = \beta v$ (with respect to ground) that can source significant current. Because of this enormous available gain β, with clever feedback an op amp can perform a multitude of tasks. For instance, it can perform as a linear amplifier of programmable gain as shown in Figure 10.14.

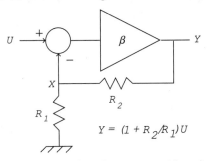

$$Y = (1 + R_2/R_1)U$$

Figure 10.14. Amplifier of programmable gain.

Since $X/Y = R_1/(R_1 + R_2)$, as in (10.13), and because $(U - X)\beta = Y$, a little algebra yields that the circuit of Figure 10.14 has transfer function

$$Y = \frac{U}{1/\beta + R_1/(R_1 + R_2)} \approx \frac{R_1 + R_2}{R_1}U, \qquad (10.24)$$

a linear amplifier of gain essentially $\alpha = 1 + R_2/R_1$. Values of say $R_1 = 10$ kΩ and $R_2 = 100$ kΩ would yield a linear amplifier of gain $\alpha = 11$.

Or for a second example, an op amp can perform as a lowpass filter, as shown in Figure 10.15. As in Figure 10.14, reprogram the gain to $\alpha = 2$. Then add a resistor/capacitor network as shown in Figure 10.15. In Exercise 10.38 you are asked to establish that the resulting closed-loop system is the lowpass filter

$$Y = \frac{2}{s^2 + s + 1}U. \qquad (10.25)$$

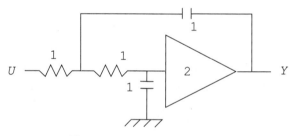

Figure 10.15. Active lowpass filter.

Example. Feedback can be used to stabilize plants. In Figure 10.16 the output $Y(s)$ of the plant $P(s)$ is fed back through the filter $H(s)$ to give $Y(s) = [U(s) - H(s)Y(s)]P(s)$; i.e., the closed-loop transfer function is

$$G(s) = \frac{Y(s)}{U(s)} = \frac{P(s)}{1 + H(s)P(s)}. \qquad (10.26)$$

The unstable plant $P(s) = 1/(s-1)$ can be stabilized by simple proportional feedback $H(s) = \kappa$ (a linear amplifier of gain κ) since once the loop is closed, the transfer function becomes

$$G(s) = \frac{1/(s-1)}{1 + \kappa/(s-1)} = \frac{1}{s + \kappa - 1},$$

which is stable for any gain $\kappa > 1$.

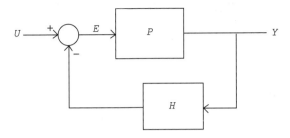

Figure 10.16. Closed-loop system.

Example. Suppose that $P(s)$ is the plant

$$P(s) = \frac{s^2 + 7s + 4}{s^3 - 3s^2 + 2s + 5}. \tag{10.27}$$

Is $P(s)$ stable? Routine 10.2 settles the matter immediately:

```
% Routine 10.2   Poles
denom = [1,-3,2,5];      % define the denominator
poles = roots(denom);    % find its roots
```

This routine returns the roots of the denominator: $s = 1.9521 \pm 1.3112i$ and -0.9042. The plant (10.27) is unstable since the conjugate pole pair lies in the right half plane.

Let us see if we can stabilize this plant $P(s)$ with negative proportional feedback, with $H(s) = \kappa$ in Figure 10.16. According to (10.26), the closed loop has transfer function

$$Q(s) = \frac{(s^2 + 7s + 4)/(s^3 - 3s^2 + 2s + 5)}{1 + \kappa(s^2 + 7s + 4)/(s^3 - 3s^2 + 2s + 5)}$$

$$= \frac{s^2 + 7s + 4}{s^3 - 3s^2 + 2s + 5 + \kappa(s^2 + 7s + 4)}. \tag{10.28}$$

The closed-loop plant (10.28) is stable if and only if its poles, i.e., the roots of

$$s^3 - 3s^2 + 2s + 5 + \kappa(s^2 + 7s + 4) = 0 \tag{10.29}$$

all lie in the open left hand plane Re $s < 0$. Let us slowly dial up feedback gain κ. As κ increases, the poles (the roots of (10.29)) will themselves move away from the open-loop poles and migrate toward the open-loop zeros (Exercise 10.40). Will they all eventually move left into the open left half plane? Routine 10.3 will perform this *root locus*:

```
% Routine 10.3  Root locus
kmin=0;                                % minimum gain
kmax = 5;                              % maximum gain
N=10;                                  % number of gain steps
dk = (kmax-kmin)/N;                    % gain step size
k = kmin:dk:kmax;                      % vector of gains to try
for n=1:N+1                            % rachet up gain
  denom = [1,-3,2,5]+k(n)*[0,1,7,4];   % denominator
  poles = roots(denom);                % poles as column vector
  re = real(poles);                    % poles real parts
  im = imag(poles);                    % poles imag. parts
  if n==1
    a=re;                              % open loop real parts
    b= im;                             % open loop imag. parts
  else
    a = [a;re];                        % stack up real parts
    b = [b;im];                        % stack up imag. parts
  end;                                 % end if/else
end;                                   % end loop
plot(a,b,'x')                          % migrating poles
```

This routine ratchets up feedback gain κ from $\kappa = 0$ to $\kappa = 5$ in 10 equal steps to yield the root-locus plot of Figure 10.17. By the eighth step, when $\kappa = 4$, the open-loop unstable pole pair $s = 1.9521 \pm 1.3112i$ has migrated to the stable pair $-0.1476 \pm 5.4562i$ while the open loop stable real pole -0.9042 has moved to the still stable -0.7049. The closed-loop system becomes stable with gain $\kappa \approx 3.72$ (Exercise 10.41).

The root locus technique can be applied more generally to any feedback element $H(s)$. We may observe how the closed-loop poles of

$$Q(s) = \frac{P(s)}{1 + \kappa H(s)P(s)} \qquad (10.30)$$

migrate with root locus applied to $1 + \kappa H(s)P(s) = 0$. We shall do examples below.

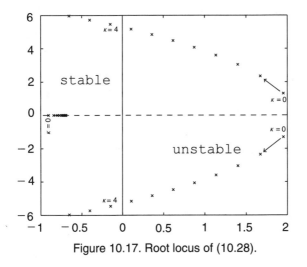

Figure 10.17. Root locus of (10.28).

*§10.8 Nyquist Analysis

There is an alternative to the root locus method that is of deep theoretical interest. The stability of (10.30) is equivalent to finding a gain κ so that the roots of

$$1/\kappa + H(s)P(s) = 0 \tag{10.31}$$

all lie in the open left half plane Re $s < 0$. The correct gains κ can be discovered via *Nyquist analysis*.

Suppose that the strictly proper rational function $G(s) = H(s)P(s)$ has unstable poles, none on the imaginary axis. Graph the image Γ of the imaginary axis $s = \omega i$ taken in the direction of increasing ω under the map $z = G(s)$. Then *all κ so that $-1/\kappa$ lies in the unbounded component of the z-plane disjointed by Γ will destabilize the closed loop* (10.30). The κ that are guaranteed to *stabilize* the closed loop are found by a more delicate analysis outlined below.

Example. Take $P(s) = 1/(s-1)$ and $H(s) = 1/(s-2)$. Is there some gain $\kappa > 0$ so that

$$Q(s) = \frac{P(s)}{1 + \kappa H(s)P(s)} = \frac{1}{s^2 - 3s + 2 + \kappa} \tag{10.32}$$

becomes stable?[1] The Nyquist plot Γ of

$$G(s) = H(s)P(s)$$

is displayed in Figure 10.18. Note that the entire negative real axis is outside the curve — thus there is no $\kappa > 0$ that stabilizes (10.32).

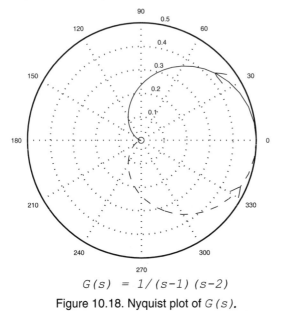

$$G(s) \;=\; 1/(s-1)(s-2)$$

Figure 10.18. Nyquist plot of $G(s)$.

The portion of the Nyquist plot Γ for negative ω is of course the complex conjugate of the portion for positive ω and is shown as a dashed curve. Figure 10.18 was obtained with Routine 10.4 below.

Example. Let us redo our earlier example of proportional feedback (10.28). Consider the unstable plant $P(s) = (s^2 + 7s + 4)/(s^3 - 3s^2 + 2s + 5)$ and $H(s) = 1$. Let us draw the Nyquist plot of

$$G(s) = H(s)P(s) = \frac{s^2 + 7s + 4}{s^3 - 3s^2 + 2s + 5} \qquad (10.33)$$

for $s = \omega i$.

```
% Routine 10.4  Nyquist plot
W=100;                              % graph out to this freq.
dw = 0.1;                          % in steps of this size
```

---

[1]The answer is clearly no, since such second-order plants are stable exactly when their coefficients are all of the same sign.

```
w=0:dw:W;                             % vector of freqs.
s= i*w;                               % along imag.  axis
N = 1;                                % numerator of H(s)
D = 1;                                % denominator of H(s)
H = N./D;                             % H(s)
N = s.*s + 7*s +4;                    % numerator of P(s)
D = s.*s.*s - 3*s.*s + 2 s +5;        % denomator of P(s)
P= N./D;                              % P(s)
G= H.*P;                              % G(s) = H(s)P(s)
r = abs(G);                           % modulus of G(s)
theta = angle(G);                     % arg of G(s)
polar(theta,r);                       % plot for positive freqs.
hold;                                 %
polar(-theta,r,'--');                 % plot for neg.  freqs.
```

This routine generates the Nyquist plot Γ of Figure 10.19. As you can observe, the entire negative real axis is outside Γ except for the portion within the inner loop from approximately $-1/4$ to 0. Thus to stabilize, $-1/\kappa$ must fall within this inner loop, i.e., $\kappa > 4 \pm \epsilon$, agreeing with our root-locus finding in Figure 10.17. As seen below, these κ do in fact stabilize the closed loop.

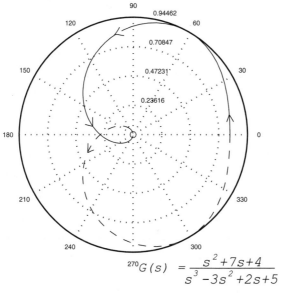

$$G(s) = \frac{s^2+7s+4}{s^3-3s^2+2s+5}$$

Figure 10.19. Nyquist plot of $G(s)$.

Justification of the Nyquist Method. Recall the *argument principle* of complex analysis [Ahlfors].

THEOREM. Suppose $f(s)$ is meromorphic on a domain Ω containing a simple closed contour C that does not pass through any zero or pole of $f(s)$. Then

$$\frac{1}{2\pi i} \int_C \frac{f'(s)}{f(s)} ds = h - k, \tag{10.34}$$

where h and k are the properly counted number of zeros and poles enclosed by C.

PROOF. For any zero or pole s_0 within C, we may write $f(s) = (s - s_0)^m g(s)$, where g(s) is analytic and nonzero at s_0. Since

$$\frac{f'(s)}{f(s)} = \frac{m}{s - s_0} + \frac{g'(s)}{g(s)},$$

$$\frac{1}{2\pi i} \int_C \frac{f'(s)}{f(s)} ds = m + \frac{1}{2\pi i} \int_C \frac{g'(s)}{g(s)} ds. \tag{10.35}$$

Apply the same argument to $g(s)$, and so on. Note that this computation assumes that the contour C is taken in the positive (counterclockwise) direction.

COROLLARY. Let Γ be the image of the contour C under $z = f(s)$, and let z_0 be a point not on Γ. Then the *winding number*

$$n(z_0, \Gamma) = \frac{1}{2\pi i} \int_\Gamma \frac{dz}{z - z_0} = \frac{1}{2\pi i} \int_C \frac{f'(s)\, ds}{f(s) - z_0} = h(z_0) - k. \tag{10.36}$$

Thus the number of windings $n(z_0, \Gamma)$ of the image contour Γ about z_0 counts the number of roots $h(z_0)$ of $f(s) = z_0$ less the number k of poles of $f(s)$ within C.

The curve Γ divides the z-plane into a finite number of (connected) components where the winding number is integral and constant. For example, $n(z_0, \Gamma) = 0$ everywhere on the unbounded component of Γ. Or for another example, $n(z_0, \Gamma) = h - k$ on the component containing 0.

H. Nyquist of Bell Laboratories cleverly applied the argument principle to stability problems by taking as the contour C what is today called the *Nyquist contour* — the imaginary axis is taken from $-Ri$ to Ri, then completed to a closed curve by the semicircle $s = Re^{i\theta}$ as θ runs from $\pi/2$ to $-\pi/2$. The radius R is taken to be very large so that in the limit the contour C will include the entire closed right half complex plane (see Figure 10.20).

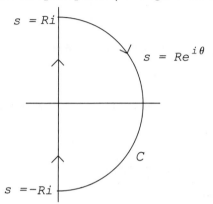

Figure 10.20. Nyquist contour.

Unfortunately this curve is oriented *in the negative direction*, (clockwise), creating no end of miscommunication between mathematicians and engineers. Keep in mind that to the engineer, the argument principle reads *poles minus zeros* rather than 'zeros minus poles.'

So to check the stability of a proper rational plant

$$P(s) = \frac{N(s)}{D(s)},$$

we must see that no poles lie within the Nyquist contour C of Figure 10.20 for all large R. We graph the *Nyquist plot* Γ, the image of C under $z = P(s)$. Suppose we know the number h of right half plane zeros of the numerator $N(s)$ from say the Nyquist plot of $1/N(s)$. Then since the winding number $n(0, \Gamma) = p - h$ is clear by inspection, the number of unstable poles p of $P(s)$ is now known: if $p = 0$, the plant $P(s)$ is stable.

Example. Let

$$P(s) = \frac{s^2 - 2s + 1}{s^3 - s^2 + s + 3}. \tag{10.37}$$

The numerator $N(s) = (s-1)^2$ has $h = 2$ zeros at $s = 1$ in the right hand plane. Graph the Nyquist plot Γ under $z = P(s)$ using Routine 10.4 (Exercise 10.44). The portion of Γ coming from the semicircle $s = Re^{i\theta}$ for large R is

$$z = P(Re^{i\theta}) \approx (Re^{i\theta})^2/(Re^{i\theta})^3 = (1/R)e^{-i\theta} \qquad (10.38)$$

and so 0 lies in the unbounded component of Γ giving

$$0 = n(0, \Gamma) = p - 2, \qquad (10.39)$$

so $P(s)$ possesses two unstable poles.

Example. Nyquist analysis is more often used to decide the stability of a closed-loop system

$$Q(s) = \frac{P(s)}{1 + \kappa H(s)P(s)},$$

as in (10.33). In this application, stability is determined by the location of the *zeros* of $1/\kappa + H(s)P(s)$, not its poles. For after all we know that in this example the number of unstable poles is constantly $p = 2$, namely the conjugate pair of unstable poles of $P(s)$. From Figure 10.19, Γ winds twice about points of the innermost component. Thus for κ such that $-1/\kappa$ lands in this innermost loop, $2 = n(-1/\kappa, \Gamma) = 2 - h$, so $h = 0$, i.e., $1/\kappa + H(s)P(s) \neq 0$ on the right half plane, so the closed loop (10.33) is stable for $\kappa > 4$.

§10.9 Control

Problem. A servomechanism is constructed from a plant $P(s) = 1/s$ with *proportional-integral* (PI) controller $H(s) = \kappa + \alpha/s$ as in Figure 10.21. Find gains κ and α that yield a step response with at most 20% overshoot and a 1-second rise time.

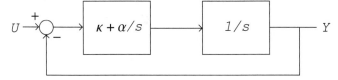

Figure 10.21. Servo with PI control.

Solution. A *selsyn* is an example of such a system, where turning a shaft through an angle θ is repeated at a remote location with amplified torque. The ubiquitous electronic *phase-locked loop* (PLL) is another example of a servomechanism (but with a much faster rise time) [Gardner]. The closed-loop transfer function from Figure 10.21 is

$$Q(s) = \frac{Y(s)}{U(s)} = \frac{H(s)P(s)}{1 + H(s)P(s)} = \frac{\kappa s + \alpha}{s^2 + \kappa s + \alpha}. \tag{10.40}$$

The step response, the response to the Heaviside unit step $U(s) = 1/s$, is

$$Y(s) = \frac{\kappa s + \alpha}{s^2 + \kappa s + \alpha} \frac{1}{s} = \frac{1}{s} - \frac{s}{s^2 + \kappa s + \alpha}. \tag{10.41}$$

Going over to standard notation (9.28) with $\kappa = 2\zeta\omega_n$, $\alpha = \omega_n^2$, and $\omega_d = \sqrt{1 - \zeta^2}\omega_n$, the step response (10.41) becomes

$$Y(s) = \frac{1}{s} - \frac{s}{s^2 + 2\zeta\omega_n s + \omega_n^2}$$

$$= \frac{1}{s} - \frac{s + \zeta\omega_n}{(s + \zeta\omega_n)^2 + \omega_d^2} + \frac{\zeta}{\sqrt{1 - \zeta^2}} \frac{\omega_d}{(s + \zeta\omega_n)^2 + \omega_d^2}, \tag{10.42}$$

which comes back to the time domain as

$$y(t) = 1 - e^{-\zeta\omega_n t} \cos \omega_d t + \frac{\zeta}{\sqrt{1 - \zeta^2}} e^{-\zeta\omega_n t} \sin \omega_d t$$

$$= 1 - \frac{1}{\sqrt{1 - \zeta^2}} e^{-\zeta\omega_n t} \cos(\omega_d t + \phi), \tag{10.43}$$

where $\phi = \arctan \zeta / \sqrt{1 - \zeta^2}$.

Next, we simulate (10.43) in order to select a damping ratio ζ and an undamped natural frequency ω_n that will meet the overshoot and rise time specifications of the problem. We go over to normalized time $\tau = \omega_n t$ and graph

$$y = 1 - \frac{1}{\sqrt{1 - \zeta^2}} e^{-\zeta\tau} \cos(\tau\sqrt{1 - \zeta^2} + \phi) \tag{10.44}$$

for various damping ratios $\zeta = 0.1, 0.2, \dots, 0.9$ to obtain Figure 10.22 (Exercise 10.45).

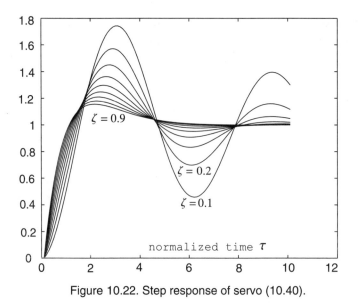

Figure 10.22. Step response of servo (10.40).

From Figure 10.22 we select $\zeta = 0.7$ to keep overshoot at or below 20%. The response for this choice of damping rises to approximately 90% of its final value of 1 by normalized time $\tau = \omega_n t = 1.5$. Since the specification for rise time is $t = 1$, we must choose undamped frequency $\omega_n = \tau/t = 1.5$. Therefore, the specifications for our PI design (10.40) require proportional gain $\kappa = 2\zeta\omega_n = 2.1$ and integral gain $\alpha = \omega_n^2 = 2.25$.

Problem. Design an automobile cruise control system.

Solution. This is a reprise of the earlier problem of §3.7. We have a closed loop that must return the vehicle speed back to set speed in the face of disturbances (see Figure 10.23). The transfer function of speed error E to load disturbance U is (Exercise 3.24)

$$\frac{E(s)}{U(s)} = \frac{-1}{1 + H(s)P(s)}. \tag{10.45}$$

Let us think about modeling the motive plant $P(s)$. The speed of the vehicle is certainly not simply proportional to the angle of depression θ of the accelerator — there is significant delay in reaching a new operating point after a sudden change in accelerator position. There are *dynamics* involved, i.e., derivatives. Acceleration $\dot{y}$ must depend directly on engine torque τ restrained by a load that increases with speed y: as a first cut,

$$M\dot{y} = -ky + K\tau.$$

Engine torque τ must depend dynamically on accelerator pedal angle θ. There is delay as the engine responds to a call for torque because of the natural regenerative nature of the internal combustion engine — a greater volume rate of fuel/air can be drawn in only if the rpm is high, and conversely.

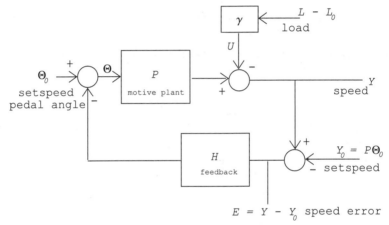

Figure 10.23. Cruise control system.

So as a first cut, let us try the first order

$$\dot{\tau} = -c\tau + d\theta.$$

Cascading, we obtain a proposed model, overdamped harmonic motion

$$P(s) = \frac{\alpha}{(s+a)(s+b)} \tag{10.46},$$

where, say, $0 < a < b$. All this is local behavior, the result of linearizing about some operating point, say 60 mph. This model seems to capture familiar behavior — consider a step change in θ. Taking $a = 1$ and $\alpha = b = 2$ gives the step response of Figure 10.24.

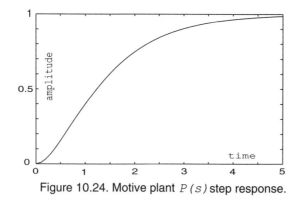

Figure 10.24. Motive plant $P(s)$ step response.

Will proportional feedback $H(s) = \kappa$ cruise control yield acceptable performance? When the loop is closed as in (10.45) and Figure 10.23, will speed return to set speed when load is perturbed by hills and wind gusts? The closed-loop step response (10.45) is

$$E(s) = -\frac{(s+a)(s+b)}{s(s^2 + (a+b)s + \alpha\kappa + ab)}$$

$$= \frac{-1}{(1 + \alpha\kappa/ab)s} + \frac{As + B}{s^2 + (a+b)s + \alpha\kappa + ab} \qquad (10.47)$$

with "final value" (*dc offset*) at $t = \infty$, after transients have died away, of

$$e_\infty = \frac{-1}{1 + \alpha\kappa/ab}. \qquad (10.48)$$

By choosing feedback gain κ large, the error e_∞ from setpoint can be brought to acceptable levels.

But what effect has large feedback gain κ on transient response? The transient term in (10.47) is (exercise),

$$\frac{As + B}{s^2 + (a+b)s + \alpha\kappa + ab} = \kappa\alpha\frac{s/\omega_n^2 + 2\zeta/\omega_n}{s^2 + 2\zeta\omega_n s + \omega_n^2}. \qquad (10.49)$$

As κ is increased, the natural undamped frequency $\omega_n = \sqrt{ab + \kappa\alpha}$ increases, forcing the damping ratio $\zeta = (a+b)/2\omega_n$ small, which leads to overshoot, as, for example, in Figure 10.22. Large overshoot is unacceptable. Excessive rise time is also unacceptable. A simulation study is warranted, which is given as a student project (Exercise 10.50).

Problem. Analyze the pitch stability of the C.S.S. *Hunley*.

Solution. The Confederate States Ship *Horace L. Hunley* was the first submarine to sink another warship in combat [Cussler and Dirgo]. The sinking took place during the U.S. blockade of Charleston harbor at 8:45 p.m. on February 17, 1864. The hapless victim was the 1240-ton sloop-of-war U.S.S. *Housatonic*. The *Hunley* rammed the *Housatonic* with a detachable explosive barb, backed away paying out a line, then ignited the 100 pounds of black-powder explosive with a yank on the line. Within 5 minutes the *Housatonic* had sunk. The *Hunley* was (possibly) inadvertently run down by the U.S.S. *Canandaigua* steaming to the rescue of the crew of the *Housatonic* with the loss of all nine submariners.

The *Hunley* had had an unfortunate run of luck — killing several crews. Although the *Hunley* anticipated many modern design inovations, it may have suffered from a basic pitch instability (see Figure 10.25).

Figure 10.25. C.S.S. *Hunley*.

The submarine would submerge and trim by letting water into the ballast tanks separately at bow and stern. The craft would surface by expelling the ballast water with two hand-cranked positive-displacement pumps. Unfortunately, both ballast compartment bulkheads were open at the top to the interior of the boat (see Figure 10.25), a flaw shared with the *Titanic*.

Water pressure increases in proportion to depth. Neutral buoyancy is achieved when the weight of the craft is exactly balanced by the net force from the water pressure on the surface area of the craft.

Once pitch reaches an extreme angle, allowing water to pour over the high-end compartment bulkhead, all is lost. But for moderate angles, is the craft inherently stable or unstable in pitch? The pitch stability of the *Hunley* is given as a project in Exercise 10.51.

Exercises

10.1 Operationally solve $\ddot{x} + x = \cos 2t$ subject to $x(0) = \dot{x}(0) = 0$.

10.2 Operationally solve $\ddot{x} + x = \cos t$ subject to $x(0) = \dot{x}(0) = 0$. Show that destructive resonance occurs — oscillations grow without bound.

10.3 Establish (10.4b).

10.4 Verify the partial fraction expansion (10.6).

10.5 Operationally solve the system $\dot{x} = x - 4y$, $\dot{y} = 3x - 2y$ subject to $x(0) = 2$ and $y(0) = -1$.

10.6 Prove there is no ordinary signal $d(t)$ such that $d(t) * f(t) = f(t)$ for some nonzero ordinary signal $f(t)$; i.e., prove that the Dirac δ is not an ordinary function.

10.7 Demonstrate by examples that integration by parts in the time domain becomes partial fraction decomposition in the frequency domain.

10.8 Deduce (10.9).

10.9 Operationally solve the system $\dot{x} = x - 2y$ and $\dot{y} = 2x - 3y$ for any initial conditions $x(0), y(0)$.

10.10 Find the time-domain version of the cascaded plant of Figure 10.1.

10.11 Solve the counterflow heat exchanger (9.37) operationally.

10.12 What is the eventual minimum value of y governed by the system $\tau \dot{y} + y = u$ with *time constant* τ when u is the unit periodic square wave $u(t) = [1 + \text{sgn}(\sin 2\pi t/T)]/2$ of period T?

Answer: $1/(1 + e^{T/2\tau})$.

10.13 Show operationally that the oscillations of a driven damped harmonic oscillator $m\ddot{x} + c\dot{x} + kx = a \cos \omega_0 t$, where $m, c, k > 0$, may grow large but remain bounded, even when driven at its resonant frequency $\omega_0 = \omega_d$. See (9.28).

10.14 Verify the rule for combining impedances in series and parallel shown in Figure 10.6.

Hints: For the series circuit, the current I flows through both impedances; temporarily introduce a symbol W for the voltage between Z_1 and Z_2. For the parallel circuit, the current I divides into two currents, $I = I_1 + I_2$.

10.15 Show that a filter constructed from resistors, capacitors, and inductors has a *strictly proper rational* transfer function (10.14), i.e., $m < n$. Show that the filter is stable as long as at least one resistor is used somewhere in the circuit (other than in the final output leg).

10.16 Prove that the transfer function of the network shown in Figure 10.8 is given by (10.13). As an epigram, *the voltage drop Y is to the whole voltage drop U as its impedance is to the whole impedance.*

10.17 Using (10.13), find the transfer function of the peak filter

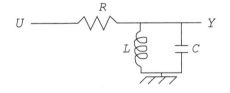

10.18*Supply all the details in the proof of Theorem A in §10.5 on the stability of proper rational plants.

10.19 Deduce Theorem A from Theorem B in §10.5.

10.20 The real parts of the roots of $3s^2 + 2s + 5 = 0$ are negative. Show that this generalizes to any quadratic polynomial with all three coefficients positive.

10.21 Show that $\exp(-t^2)$ is not a rational function of generalized differentiation s.

10.22 Show that $1/\sqrt{t}$ is not a rational function of s.

10.23*Show $\{t^\alpha\} = \Gamma(\alpha + 1)/s^{\alpha+1}$ for all real numbers $\alpha > -1$.

10.24 Modify Routine 10.1 to graph the *polar plot* of the notch filter (10.19). The polar plot is the graph in the complex plane of $z = r(w)e^{i\phi(w)}$ (see Routine 3.1).

10.25*Using the ideas of the proof of Theorem B, §10.5, work Exercise 3.15 completely.

10.26**We say that a plant $y = p * u$ given by convolution with an ordinary signal $p(t)$ is *stable in energy* when ordinary square-summable inputs yield ordinary square-summable outputs; i.e., $\|u\|_2 < \infty \Rightarrow \|y\|_2 < \infty$. Show that convolution by p is energy stable if and only if p is in H^∞; i.e., the Fourier transform of p is a bounded function.

10.27**Formulate and prove a criterion for stability in energy for discrete-time filters.

10.28 Draw the Bode plots of the peak filter of Exercise 10.17 when $L = C = 1$ and $R = 1000$.

10.29 Show that the filter given by

$$H(s) = \frac{s/Q}{s^2 + s/Q + 1}$$

peaks in response at $w = 1$ with gain 1, and that the phase lag is more or less constantly $\pm 90°$, more so with increasing Q. Moreover, establish that this figure of merit Q is in fact the center frequency divided by the *bandwidth*, i.e., the width between the -3-dB points.

10.30 Why is it important that a bandpass filter have *constant group delay* throughout its passband, i.e., that the phase lead/lag ϕ in (3.40b) and (10.17b) be close to a linear function of frequency w?

Hint: Simultaneous zero crossings of signals in the passband must be preserved to prevent *phase distortion*.

10.31 Modify Routine 10.1 to obtain the Bode plots of the bandpass filter shown in Figure 10.11.

10.32 Once you have in hand the routine for the Exercise 10.31, experiment with the filter (10.22) by moving the center frequency of $L(s)$ and $R(s)$ to see if performance can be improved. Try higher Q values. Try cascading a third filter of type (10.21) centered between 10 and 15 Hz.

10.33 The *nth-order Butterworth lowpass filter* is formed by using the $2n$th roots of unity $x^{2n} - 1 = 0$ that lie in the open left half plane:

$$F(s) = \frac{1}{D(s)},$$

where $D(s) = \prod(s - \xi)$, where $\xi^{2n} = 1$ and Re $\xi < 0$. Find the third-order Butterworth LPF $F(s)$ and its Bode gain plot. Rescale frequency so that the -3-dB point is at $f = 3$ kHz.

10.34 Obtain the transfer function (9.48) of following to leading vehicle in traffic.

10.35 Is the filter $y(t) = u(t/2)$ linear, causal, and time invariant?

10.36 Note that the notch filter $N(s)$ of (10.19) is related to the peak filter $P(s)$ of (10.20) when $Q = 1$ by the relation $1 - P(s) = N(s)$. Is this accidental? What about a similar relation between lowpass and highpass filters?

10.37 For proper rational transfer functions, show that the substitution $s \to 1/s$ transforms stable lowpass filters $F(s)$ into stable highpass filters.

10.38 Establish (10.25).

Outline: The current leaving the input source U splits into two currents at the junction of the two resistors and capacitor. Write an equation for the current through, and voltage across, each component.

10.39 At what gain α for the linear amplifier in Figure 10.15 does the system become unstable and oscillate? What is its frequency ω of oscillation?

Answer: $\alpha = 3$, $\omega = 1$.

10.40 Show that as the feedback gain κ increases, the poles of the closed-loop transfer function $G(s)$ of (10.28) migrate from the open-loop poles [of $P(s)$] to either ∞ or to the open-loop zeros [of $P(s)$].

10.41 Experiment with Routine 10.3 to find the gain κ at which the closed loop (10.28) becomes stable.

10.42 Modify Routine 10.3 to find if there is a gain κ that stabilizes $P(s) = (s^2 + s + 1)/(s^4 + s^3 - 5s^2 - s - 1)$ using proportional feedback $H(s) = \kappa$ as in Figure 10.16. Add on several lines that will add to the plot the open-loop zeros (marked with a "o"). Note how some of the closed-loop poles migrate towards the open-loop zeros, as guaranteed by Exercise 10.40.

10.43 Generalize Routine 10.3 to perform a root-locus procedure on the general case $1 + \kappa H(s)P(s)$ where $H(s)$ and $P(s)$ are rational. Apply to $H(s) = 1/(s + 1)$ and $P(s)$ given in (10.27).

10.44 Obtain the Nyquist plot of (10.37) using Routine 10.4.

10.45 Write a routine to obtain the graphs of Figure 10.22.

10.46 Use the Nyquist plot to analyze the stability of $P(s) = (s^3 - 2s + 4)/(s^4 - 2s^3 + 4s^2 - 5s + 2)$. Check your result with Routine 10.2.

10.47 Redesign the servo of Figure 10.21 for at most 15% overshoot but a rise time of 1 ms.

10.48*(Project) Obtain the stability condition $\tau\alpha < \pi/2$ for the two-vehicle traffic problem (9.48). Show that even when stability is satisfied, a long platoon of vehicles is inherently unsafe since a change in the lead vehicle speed produces larger and larger excursions further down the line.

10.49 (Project) Research the four classical approaches to control design: Nyquist, root-locus, Bode, and Nichols chart, the pantheon of "Bell Lab Mathematics." Compare their effectiveness on simple example control problems.

10.50 (Project) Simulate proportional feedback cruise control, (10.45)–(10.49). Using a stopwatch and a passenger car, determine realworld values for the poles a and b. Then simulate performance for various gains κ. Is there an acceptable compromise yielding low error e from setpoint, reasonable rise time, and a small overshoot to a step input? Is a more complicated PID controller $H(s) = \kappa + \lambda/s + \mu s$ warranted?

10.51 (Project) Analyze the pitch stability of the submarine C.S.S. *Horace L. Hunley.* If the craft is perturbed off an even keel, are there forces tending to restore trim or to increase pitch? If unstable, propose remediation consonant with the technology of the time.

10.52* It is a famous result of feedforward control theory that any reachable state $x(T)$ of an initially quiescent system $\dot{x} = Ax + Bu$ obtainable by a control $u = u(t)$, $0 \le t \le T$, with $0 \le u(t) \le 1$, can also be reached at time $t = T$ by *bang-bang* control, by a control $u(t)$ on $0 \le t \le T$ that takes on only the two values $0, 1$. Think of attitude thrusters on spacecraft. Prove this result for the scalar system $\dot{x} = -2x + u$.

10.53 (Project) Investigate the advisability of stabilizing the roll of a cruise ship by pumping ballast water from a central tank below the center of gravity to the high side in order to return the ship to vertical. Comment on the rule of thumb that *excessive delay in a feedback loop is dangerous.*

10.54* Prove *Bode's theorem:* The phase lead $\phi = \arg H(i\omega)$ of a strictly proper stable rational filter $H(s)$ with no right-half-plane zeros (i.e., of *minimum phase*) can be reconstructed from its gain $r = |H(i\omega)|$.

Outline: After drawing a small circular left indentation about $i\omega_0$, a single-valued analytic branch of $\log H(s)$ can be defined on the resulting slightly enlarged right half plane. By Cauchy,

$$\log H(i\omega_0) = \frac{-1}{2\pi i} \int_C \frac{\log H(s)}{s - i\omega_0} ds$$

where C is the Nyquist contour with the indentation. Compute the imaginary part of both sides to obtain

$$\arg H(i\omega_0) = \frac{1}{\pi} \int_{-\infty}^{\infty} \frac{\log |H(i\omega)|}{\omega - \omega_0} d\omega.$$

10.55* Deduce from (10.16) that a stable plant given by convolution with an ordinary signal $p(t)$ (an *ordinary plant*) has a transfer function $P(s)$ that is analytic on the open right half plane, continuous on its closure, with

$$\lim_{s \to \infty} P(s) = 0.$$

10.56*The transfer function of an ordinary stable plant may be free of poles altogether. Show the Laplace transform $F(s)$ of

$$f(t) = e^{-t^2}$$

is everywhere analytic, then find $F(s)$ using that $2F'(s) = sF(s) - 1$.

Answer:

$$F(s) = \sqrt{\pi} \, e^{s^2/4} - \frac{e^{s^2/4}}{2} \int_0^s e^{-z^2/4} \, dz.$$

10.57**For nonrational transfer functions, the absence of right-half-plane poles does not guarantee stability. Find an ordinary function $f(t)$ with Laplace transform $F(s)$ that is everywhere analytic on Re $s \geq 0$, yet convolution by f is not a stable plant.

10.58 Find an ordinary stable plant whose impulse response is unbounded.

Chapter 11

Partial Differential Equations

The most fundamental of all phenomena are described by PDEs.

We begin by drawing a distinction between lumped and distributed problems. Most distributed problems are modeled by one of the big six PDEs: diffusion, wave, Laplace, Poisson, Schrödinger, and the plate equation. Typical problems are worked for each type by separation of variables. Methods for unbounded spatial domains are surveyed. Helmholtz's phasor method is introduced to find periodic steady-state deep ground temperatures. Finally, the PDEs modeling acoustics, electromagnetics, transport, expressway traffic, and fluid flow are introduced.

§11.1 Lumped versus Distributed

Some systems are by their very nature infinite-dimensional. In contrast, the motion of a pendulum, mass–spring system, a satellite, a piston, or voltages in resistor/capacitor/inductor circuits, can be predicted once the values of some finite select list of scalar variables and their derivatives are known. These are called *lumped* problems and are modeled by ODEs. But other problems cannot be predicted given any finite list of scalar measurements. Think of the deflections of a long bridge span — knowing the deflections at any number of places and their derivatives in time and space will not yield a successful prediction of future shapes. Or think of transient voltages down a transmission line, or temperatures within a solid. These are *distributed* problems, requiring measurements over regions. Distributed problems do, however, resemble lumped problems in that they often can be written as a (now infinite) system of first- or second-order

191

ODEs. This method of reduction to an infinite system of ODEs is an outcome of the central method of *separation of variables* reviewed in §11.3.

§11.2 The Big Six PDEs

The most frequently encountered distributed problems are modeled by variations on one of the "Big Six" PDEs:

1. **The heat equation (or equation of diffusion)**

$$\frac{\partial u}{\partial t} = \nabla^2 u \qquad (11.1)$$

2. **The wave equation**

$$\frac{\partial^2 u}{\partial t^2} = \nabla^2 u \qquad (11.2)$$

3. **Laplace's equation**

$$\nabla^2 u = 0 \qquad (11.3)$$

4. **Poisson's equation**

$$\nabla^2 u = f \qquad (11.4)$$

5. **Schrödinger's equation**

$$i\frac{\partial \psi}{\partial t} = -\nabla^2 \psi + v(x)\psi \qquad (11.5)$$

6. **The plate (beam) equation**

$$\frac{\partial^2 u}{\partial t^2} = -\nabla^2 \nabla^2 u \qquad (11.6)$$

Some of the above have been brought to their *nondimensional* form by an appropriate rescaling of variables.

Recall that the *del* operator

$$\nabla = \frac{\partial}{\partial x}\mathbf{i} + \frac{\partial}{\partial y}\mathbf{j} + \frac{\partial}{\partial z}\mathbf{k} \qquad (11.7)$$

when applied to a scalar function f yields the *gradient*

$$\nabla f = \frac{\partial f}{\partial x}\mathbf{i} + \frac{\partial f}{\partial y}\mathbf{j} + \frac{\partial f}{\partial z}\mathbf{k}, \tag{11.8}$$

a vector field, while in contrast the *divergence* transforms a vector field

$$F = P\mathbf{i} + Q\mathbf{j} + R\mathbf{k}$$

into the scalar field

$$\nabla \circ F = \frac{\partial P}{\partial x} + \frac{\partial Q}{\partial y} + \frac{\partial R}{\partial z}. \tag{11.9}$$

The *curl*

$$\nabla \times F = \begin{vmatrix} \mathbf{i} & \mathbf{j} & \mathbf{k} \\ \frac{\partial}{\partial x} & \frac{\partial}{\partial y} & \frac{\partial}{\partial z} \\ P & Q & R \end{vmatrix} \tag{11.10}$$

transforms vector fields to vector fields.

As can be seen from the list (11.1)–(11.6), the most physically important operator is the **Laplacian**, the divergence of the gradient

$$\nabla^2 f = \nabla \circ \nabla f = \frac{\partial^2 f}{\partial x^2} + \frac{\partial^2 f}{\partial y^2} + \frac{\partial^2 f}{\partial z^2}, \tag{11.11}$$

transforming scalar fields to scalar fields.

In particular, when $f = f(x, y)$ and $g = g(x)$,

$$\nabla^2 f = \frac{\partial^2 f}{\partial x^2} + \frac{\partial^2 f}{\partial y^2} \tag{11.12}$$

and

$$\nabla^2 g = g''. \tag{11.13}$$

The **Laplacian in cylindrical coordinates** is

$$\nabla^2 u = \frac{\partial^2 u}{\partial r^2} + \frac{1}{r}\frac{\partial u}{\partial r} + \frac{1}{r^2}\frac{\partial^2 u}{\partial \theta^2} + \frac{\partial^2 u}{\partial z^2}. \tag{11.14}$$

The **Laplacian in spherical coordinates** is

$$\nabla^2 u = \frac{\partial^2 u}{\partial \rho^2} + \frac{2}{\rho}\frac{\partial u}{\partial \rho} + \frac{1}{\rho^2 \sin^2 \phi}\frac{\partial^2 u}{\partial \theta^2} + \frac{1}{\rho^2}\frac{\partial^2 u}{\partial \phi^2} + \frac{\cot \phi}{\rho^2}\frac{\partial u}{\partial \phi}. \tag{11.15}$$

§11.3 Separation of Variables

What follows is a carefully selected set of standard problems that capture much of the core of this subject. One good approach to learning this subject is to rework the following problems repeatedly. For the theoretical underpinnings, see MacCluer's 1994 *Boundary Value Problems and Orthogonal Expansions*.

Problem 1. A homogeneous isotropic rod of length 1 with curved side insulated is initially and uniformly at temperature 1. Its ends are suddenly brought in contact with sinks at temperature 0. How will the temperature profile of the rod relax to 0? In symbols,

$$\frac{\partial u}{\partial t} = \frac{\partial^2 u}{\partial x^2}, \quad 0 < x < 1, \tag{11.16}$$

subject to

(i) $u(x,0) = 1, \quad 0 < x < 1$ (initial condition)

(ii) $u(0,t) = 0, \quad t \geq 0$ (left boundary condition)

(iii) $u(1,t) = 0, \quad t \geq 0$ (right boundary condition)

Solution by Separation of Variables. We find special solutions satisfying only the boundary conditions (ii) and (iii), then superimpose these special solutions to realize the initial condition (i).

How special? Assume *variables separate*:

$$u(x,t) = T(t)X(x).$$

But then, if u is such a separated solution, the heat equation (11.16) becomes

$$\dot{T}(t)X(x) = T(t)X''(x),$$

where the dot is Newton's notation for differentiation with respect to time t and the prime is, as usual, differentiation with respect to the spatial variable x. Rearranging, we obtain

$$\frac{\dot{T}(t)}{T(t)} = \frac{X''(x)}{X(x)}. \tag{11.17}$$

Since the left-hand side of (11.17) is a function only of t and the right-hand side only of x, the only conclusion possible is that both members are constant, i.e.,

$$\frac{\dot{T}(t)}{T(t)} = \frac{X''(x)}{X(x)} = \lambda. \tag{11.18}$$

The constant λ is called a *separation constant* or an *eigenvalue* of the Laplacian since $\nabla^2 X = X'' = \lambda X$. The function $X(x)$ is called an *eigenfunction* of the Laplacian belonging to the eigenvalue λ.

The Laplacian with zero boundary conditions is always negative definite (Exercise 11.6), so λ must be a negative real number (Exercise 11.7), i.e.,

$$\lambda = -\omega^2. \tag{11.19}$$

Thus we are led to the decoupled ODEs

$$\dot{T} + \omega^2 T = 0$$

$$X'' + \omega^2 X = 0$$

with solutions

$$T(t) = e^{-\omega^2 t}$$

and

$$X(x) = a\cos\omega x + b\sin\omega x.$$

Imposing boundary condition (ii) on u, hence X, yields that $a = 0$. Imposing boundary condition (iii) on u, hence X, yields that $\sin\omega = 0$, so $\omega = n\pi$. Hence

$$T_n(t) = e^{-n^2\pi^2 t}$$
$$X_n(x) = \sin n\pi x$$
$$\lambda_n = -n^2\pi^2,$$

yielding our special solutions

$$u_n = e^{-n^2\pi^2 t}\sin n\pi x, \tag{11.20}$$

one for each $n = 1, 2, 3, \ldots$. But then series of the form

$$u(x,t) = \sum_{n=1}^{\infty} c_n e^{-n^2\pi^2 t}\sin n\pi x \tag{11.21}$$

will also satisfy the heat equation (11.16) and boundary conditions (ii) and (iii). Only the initial condition (i) remains to be satisfied.

Setting $t = 0$ and imposing on (11.21) the initial condition (i) yields

$$1 = \sum_{n=1}^{\infty} c_n \sin n\pi x. \tag{11.22}$$

To compute these constants c_n we employ the inner product belonging naturally to the spatial domain of this problem, namely the inner product

$$\langle f, g \rangle = \int_0^1 f(y)g(y)\,dy. \tag{11.23}$$

Apply the linear functional $\langle \sin m\pi x, \cdot \rangle$ to both sides of (11.22) to obtain

$$\langle \sin m\pi y, 1 \rangle = \sum_{n=1}^{\infty} c_n \langle \sin m\pi y, \sin n\pi y \rangle. \tag{11.24}$$

But because the Laplacian with zero boundary conditions is self-adjoint (Exercise 11.8), the functions $\sin n\pi x$ are *mutually orthogonal* (Exercise 11.9), i.e.,

$$\langle \sin m\pi y \ \sin n\pi y \rangle = \begin{cases} 0 & \text{if } m \neq n \\ \dfrac{1}{2} & \text{if } m = n. \end{cases} \tag{11.25}$$

Performing the inner products yields (Exercise 11.11)

$$c_n = \frac{\langle \sin n\pi y, 1 \rangle}{\langle \sin n\pi y, \sin n\pi y \rangle} = \frac{2}{n\pi}[1 - (-1)^n] \tag{11.26}$$

$$= \begin{cases} \dfrac{4}{n\pi} & \text{if } n \text{ odd} \\ 0 & \text{if } n \text{ even}. \end{cases}$$

Thus it is plausible that the solution to the original problem is

$$u(x, t) = \frac{4}{\pi} \sum_{k=0}^{\infty} e^{-(2k+1)^2\pi^2 t} \frac{\sin(2k+1)\pi x}{2k+1}. \tag{11.27}$$

Remark 1. It is a deep result of Franz Rellich that the eigenfunctions of the Laplacian subject to zero boundary conditions on a bounded domain Ω form a complete orthogonal basis of $L^2(\Omega)$. Thus the initial condition (11.22) is realizable in the L^2 sense (see [MacCluer, 1994]).

Remark 2. In what sense is the equality (11.27)? Is it pointwise convergence everywhere, almost everywhere, uniformly, L^2? By a deep result of Herman Weyl, the equality in the solution of the heat equation (11.27) is in the L^2 sense whenever the initial temperature $u(x, 0)$ is square-integrable, i.e., in $L^2(0, 1)$ (see [MacCluer, 1994]).

Remark 3. Suppose the initial temperature of the rod is a more complicated function $u(x, 0) = f(x)$ of location x. The calculations above remain valid until step (11.22), where, instead,

$$f(x) = \sum_{n=1}^{\infty} c_n \sin n\pi x, \tag{11.28}$$

thus yielding in the same way the general constants

$$c_n = \frac{\langle \sin n\pi y, f(y) \rangle}{\langle \sin n\pi y, \sin n\pi y \rangle}. \tag{11.29}$$

But now think of this problem as an input–output problem, where the input is the initial condition $f(x)$, the output the solution $u = u(x, t)$. Set $\phi_n(x) = \sin n\pi x / \| \sin n\pi x \| = \sqrt{2} \sin n\pi x$. Then

$$u(x, t) = \sum_{n=1}^{\infty} e^{\lambda_n t} \langle \phi_n(y), f(y) \rangle \phi_n(x)$$

$$= \langle \sum_{n=1}^{\infty} e^{\lambda_n t} \phi_n(y) \phi_n(x), f(y) \rangle = \int_0^1 G(x, y, t) f(y) \, dy, \tag{11.30}$$

where G is the *Green's function* for this problem with zero boundary conditions where

$$G(x, y, t) = \sum_{n=1}^{\infty} e^{\lambda_n t} \phi_n(y) \phi_n(x)$$

$$= 2 \sum_{n=1}^{\infty} e^{-n^2 \pi^2 t} \sin n\pi x \sin n\pi y. \tag{11.31}$$

Remark 4. What if the original problem (11.16) were not given with zero boundary conditions but instead with (ii)' $u(0,t) = a(t)$, $t \geq 0$ (left boundary condition)

(iii)' $u(1,t) = b(t)$, $t \geq 0$ (right boundary condition)?

The strategy now becomes to reduce the problem to one with zero boundary conditions with a substitution like

$$v(x,t) = u(x,t) - (1-x)a(t) - xb(t), \qquad (11.32)$$

so that the new problem becomes the nonhomogeneous problem

$$\frac{\partial v}{\partial t} = \frac{\partial^2 v}{\partial x^2} + q(x,t), \quad 0 < x < 1 \qquad (11.33)$$

subject to zero boundary conditions and some initial condition $v(x,0) = f(x)$.

Remark 5. There are two approaches to solving a nonhomogeneous diffusion problem such as (11.33).

Method 1. Use the orthonormal series expansion obtained from the homogeneous problem but with undetermined time-varying coefficients $c_n(t)$:

$$v = \sum_{n=1}^{\infty} c_n(t)\phi_n(x), \qquad (11.34)$$

where in this case $\phi_n(x) = \sqrt{2}\sin n\pi x$. Second, solve for constants f_n that expand the initial condition

$$f(x) = \sum_{n=1}^{\infty} f_n\phi_n(x). \qquad (11.35)$$

Third, find the $q_n(t) = \langle q(y,t), \phi_n(y)\rangle$ so that

$$q(x,t) = \sum_{n=1}^{\infty} q_n(t)\phi_n(x). \qquad (11.36)$$

Then by applying the PDE (11.33) term by term to (11.34) we have reduced our problem to solving the infinitely many ODEs

$$\dot{c}_n - \lambda_n c_n = q_n \qquad (11.37)$$

subject to $c_n(0) = f_n$.

Method 2. Try to throw the nonhomogenity onto the initial condition by subtracting off a particular solution with zero boundary conditions:

$$w(x,t) = v(x,t) - h(x,t), \tag{11.38}$$

where

$$\frac{\partial h}{\partial t} = \frac{\partial^2 h}{\partial x^2} + g(x,t)$$

with $h(0,t) = 0 = h(1,t)$, returning us to a homogeneous problem

$$\frac{\partial w}{\partial t} = \frac{\partial^2 w}{\partial x^2}$$

with zero boundary conditions.

Remark 6. Suppose the problem is presented dimensionally, i.e.,

$$\frac{\partial u}{\partial t} = \alpha \frac{\partial^2 u}{\partial x^2}$$

for $a < x < b$. Then make Fourier's change of time scale $\theta = \alpha t/(b - a)^2$ and Biot's change of spatial scale $\zeta = (x - a)/(b - a)$ to obtain the nondimensional

$$\frac{\partial u}{\partial \theta} = \frac{\partial^2 u}{\partial \zeta^2}$$

on $0 < \zeta < 1$.

Problem 2. Let us consider a *Robin* problem, where heat flux leaving the right end of a rod (or infinite slab) is driven by the gradient to ambient. More precisely, consider

$$\frac{\partial u}{\partial t} = \frac{\partial^2 u}{\partial x^2}, \quad 0 < x < 1, \tag{11.39}$$

subject to

(i) $u(x,0) = 1, \quad 0 < x < 1$

(ii) $u(0,t) = 0, \quad t \geq 0$

(iii) $u_x(1,t) = -u(1,t), \quad t \geq 0$

(The subscript x in boundary condition (iii) denotes partial differentiation with respect to x.)

Solution. Again assume a separable solution

$$u(x,t) = T(t)X(x)$$

leading to

$$\frac{\dot{T}(t)}{T(t)} = \frac{X''(x)}{X(x)} = \lambda = -\omega^2,$$

giving as before that

$$T(t) = e^{-\omega^2 t}$$

and after imposing (ii),

$$X(x) = \sin \omega x.$$

But in this case, imposing the right boundary condition (iii) leads instead to the transcendental

$$-\omega \cos \omega = \sin \omega,$$

i.e., the *characteristic equation*

$$\tan \omega = -\omega. \tag{11.40}$$

Graph $y = \tan \omega$ superimposed onto a graph of $y = -\omega$ as in Figure 11.1.

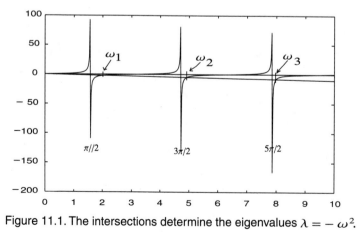

Figure 11.1. The intersections determine the eigenvalues $\lambda = -\omega^2$.

The intersections are solutions

$$0 < w_1 < w_2 < \cdots$$

to the characteristic equation (11.40), where for large n,

$$w_n \approx \frac{2n-1}{2}\pi.$$

Thus it is plausible that the solution to our problem (11.39) is of the form

$$u(x,t) = \sum_1^\infty c_n e^{-w_n^2 t} \sin w_n x. \tag{11.41}$$

Imposing the initial condition (i) leads to

$$1 = \sum_1^\infty c_n \sin w_n x. \tag{11.42}$$

Again employ the inner product (11.23) belonging naturally to this problem. Apply the linear functional $\langle \sin w_m x, \cdot \rangle$ to the initial condition (11.42) to obtain

$$\langle \sin w_m x, 1 \rangle = \sum_{n=1}^\infty c_n \langle \sin w_m x, \sin w_n x \rangle.$$

Again, because the Laplacian with these mixed boundary conditions is self-adjoint (Exercise 11.17), the functions $\sin w_n x$ are mutually orthogonal, and hence, after some computation (Exercise 11.14),

$$c_n = 2\frac{1 - \cos w_n}{w_n(1 + \cos^2 w_n)}, \tag{11.43}$$

giving a solution (11.41) to our original mixed problem (11.39).

Problem 3. (The Fourier ring problem) A long rod is bent and its ends welded to form a ring. Given an initial temperature profile around the ring, how will this temperature profile evolve over time? In symbols,

$$\frac{\partial u}{\partial t} = \frac{\partial^2 u}{\partial x^2}, \quad -\pi < x < \pi, \tag{11.44}$$

subject to the initial profile

(i) $u(x, 0) = f(x)$,

and the spatially periodic boundary conditions

(ii) $u(-\pi, t) = u(\pi, t)$,

and

(iii) $u_x(-\pi, t) = u_x(\pi, t)$.

Solution. Assuming that

$$u(x, t) = T(t)X(x)$$

yields

$$\frac{\dot{T}(t)}{T(t)} = \frac{X''(x)}{X(x)} = \lambda = -\omega^2,$$

so

$$T(t) = e^{-\omega^2 t}$$

and

$$X = a\cos\omega x + b\sin\omega x.$$

Applying boundary conditions (ii) and (iii) yields that (exercise)

$$\omega_n = n,$$

and so

$$T_n(t) = e^{-n^2 t}$$

$$X_n(x) = a_n \cos nx + b_n \sin nx$$

for $n = 0, 1, 2, \ldots$. Thus we are led to a solution of the ring problem

$$u(x, t) = \sum_{n=0}^{\infty} e^{-n^2 t}(a_n \cos nx + b_n \sin nx). \qquad (11.45)$$

The inner product belonging naturally to this problem is

$$\langle f, g \rangle = \int_{-\pi}^{\pi} f(x)g(x)\, dx. \qquad (11.46)$$

Again, since the Laplacian with periodic boundary conditions is self-adjoint (exercise), the trigonometric functions

$$1, \cos x, \sin x, \cos 2x, \sin 2x, \cos 3x, \ldots$$

are mutually orthogonal.

Applying the functionals $\langle \cos nx, \cdot \rangle$ and $\langle \sin nx, \cdot \rangle$ to the initial condition

$$f(x) = a_0 + \sum_{n=1}^{\infty} a_n \cos nx + b_n \sin nx \tag{11.47}$$

yields the famous formulas of *Fourier analysis*:

$$a_n = \frac{1}{\pi} \int_{-\pi}^{\pi} f(x) \cos nx \; dx \tag{11.48a}$$

and

$$b_n = \frac{1}{\pi} \int_{-\pi}^{\pi} f(x) \sin nx \; dx \tag{11.48b}$$

for $n > 0$, while a_0 is the average value of f, i.e.,

$$a_0 = \frac{1}{2\pi} \int_{-\pi}^{\pi} f(x) \; dx \tag{11.48c}$$

and $b_0 = 0$ (Exercise 11.15). The ring problem has led us directly to *Fourier series*, the representation of periodic functions by trigonometric series.

Problem 4. (The plucked string) Pluck a violin string and determine the vibrations produced. That is, consider the problem

$$\frac{\partial^2 u}{\partial t^2} = \frac{\partial^2 u}{\partial x^2}, \quad 0 < x < 1, \tag{11.49}$$

subject to

(i) $u(x, 0) = f(x)$

(ii) $u_t(x, 0) = 0, \quad 0 < x < 1$

(iii) $u(0, t) = 0, \quad t \geq 0$

(iv) $u(1, t) = 0, \quad t \geq 0$

Solution. As a rule of thumb, *well-posed problems have as many conditions as derivatives*. Assume a solution of the form

$$u(x, t) = T(t)X(x)$$

and impose the wave equation (11.49) to obtain

$$\ddot{T} + \omega^2 T = 0$$

and

$$X'' + \omega^2 X = 0.$$

Imposing (iii) and (iv) as in Problem 1 (11.16) yields that $\omega_n = n\pi$ and that

$$X_n(x) = \sin n\pi x.$$

Imposing initial condition (ii) on T_n yields

$$T_n(t) = \cos n\pi t$$

and hence our solution has the form

$$u(x,t) = \sum_{n=1}^{\infty} c_n \cos n\pi t \, \sin n\pi x. \qquad (11.50)$$

Physically, the spatial eigenmodes (overtones) $X_n = \sin n\pi x$ are being modulated in amplitude harmonically by the temporal modes $T_n = \cos n\pi t$.

Suppose now that the string is plucked at its center, i.e.,

$$f(x) = \begin{cases} 2x & \text{if } 0 \le x \le 1/2 \\ 2 - 2x & \text{if } 1/2 \le x \le 1. \end{cases}$$

That is, initially,

$$f(x) = \sum_{1}^{\infty} c_n \sin n\pi x$$

giving that (Exercise 11.16)

$$c_n = \frac{8 \sin(n\pi/2)}{n^2 \pi^2}. \qquad (11.51)$$

There are deep theoretical difficulties with this solution — it is a *weak* solution, not a solution in the classical sense (see Exercise 11.25).

Problem 5. (A circular drum) Let us sound a circular drum by carefully striking with the mallet at the very center so that vibration is radially symmetric. In symbols, solve

$$\frac{\partial^2 u}{\partial t^2} = \nabla^2 u, \quad 0 \le r < 1, \qquad (11.52)$$

subject to

(i) $u(r, 0) = f(r)$, $r < 1$

(ii) $u_t(r, 0) = 0$

(iii) $u(1, t) = 0$, $t \geq 0$

Solution. Although this model appears to break the rule of thumb of *one condition per derivative*, there is a latent condition that will reveal itself during the calculation. Assume

$$u(r, t) = T(t)R(r),$$

giving that

$$\frac{\ddot{T}(t)}{T(t)} = \frac{\nabla^2 R}{R} = \lambda. \tag{11.53}$$

Under the natural inner product (in polar coordinates) of this problem,

$$\langle f(r), g(r) \rangle = \int_0^1 f(r)g(r)r \, dr, \tag{11.54}$$

the Laplacian with zero boundary conditions is negative definite self-adjoint (Exercises 11.6 and 11.8). Thus we are guaranteed that the eigenvalue λ is negative and that eigenfunctions are mutually orthogonal. Set $\lambda = -\alpha^2$. Using the Laplacian in polar coordinates (11.14) on (11.53), we obtain the decoupled ODEs

$$\ddot{T} + \alpha^2 T = 0$$

and

$$\frac{d^2 R}{dr^2} + \frac{1}{r}\frac{dR}{dr} = -\alpha^2 R.$$

Multiplying the second through by r^2, then making the substitution

$$x = \alpha r, \quad y = R(r),$$

leads directly (Exercise 11.18) to *Bessel's equation:*

$$x^2 y'' + xy' + x^2 y = 0. \tag{11.55}$$

This ODE possesses two independent solutions — a bounded solution $J_0(x)$, and an unbounded solution $Y_0(x)$ near 0. Limited to bounded

solutions — our latent fourth condition — we see by the power series
method that within a constant multiple,

$$y = J_0(x) = \sum_{k=0}^{\infty} \frac{(-1)^k}{k!k!} \left(\frac{x}{2}\right)^{2k}, \tag{11.56}$$

a function known as the *Bessel function of order* 0 *of the first kind.*
Think of $J_0(x)$ as a damped version of cosine — a graph is displayed
in Figure 11.2.

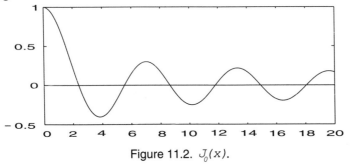

Figure 11.2. $J_0(x)$.

The far-reaching result of Rellich mentioned in Remark 1 above
guarantees that $J_0(x)$ possesses positive zeros α_n increasing to ∞:

$$0 < \alpha_1 < \alpha_2 < \alpha_3 < \cdots$$

and that the $\phi_k = J_0(\alpha_k r)$ form a complete orthogonal basis. Values
of the first six positive zeros α_k of $J_0(x)$ are as follows:

α_1	α_2	α_3	α_4	α_5	α_6
2.40483	5.52008	8.65373	11.79153	14.93092	18.07106

Returning to our problem and resubstituting $r = x/\alpha$, we obtain
the vibrations of a circular drum struck at the center:

$$u(r,t) = \sum_{n=1}^{\infty} c_n \cos \alpha_n t \, J_0(\alpha_n r). \tag{11.57}$$

Because the $J_0(\alpha_n r)$ are mutually orthogonal with respect to the
inner product (11.54),

$$c_n = \frac{\langle f(r), J_0(\alpha_n r) \rangle}{\langle J_0(\alpha_n r), J_0(\alpha_n r) \rangle} = \frac{\int_0^1 f(r) J_0(\alpha_n r) r \, dr}{\int_0^1 J_0(\alpha_n r)^2 r \, dr}. \tag{11.58}$$

Problem 6. (Quenching a ball) A solid homogeneous isotropic ball of radius 1, initially and uniformly at temperature 1, is suddenly dropped into a bath at temperature 0. Let us discover how the interior temperatures relax to 0. The solution is radially symmetric:

$$\frac{\partial u}{\partial t} = \nabla^2 u, \quad 0 \le \rho < 1, \tag{11.59}$$

subject to

(i) $u(\rho, 0) = 1, \quad 0 \le \rho < 1$

(ii) $u(1, t) = 0, \quad t \ge 0$

together with the latent condition that we search for bounded solutions.

Solution. Take the inner product belonging naturally to this problem from spherical coordinates:

$$\langle f(\rho), g(\rho) \rangle = \int_0^1 f(\rho)g(\rho)\rho^2 \, d\rho. \tag{11.60}$$

Because the boundary conditions are zero, the Laplacian is self-adjoint and negative definite. Assume that

$$u(\rho, t) = T(t)R(\rho),$$

which gives rise to

$$\frac{\dot{T}}{T} = \frac{\nabla^2 R}{R} = \lambda = -\omega^2$$

and thus

$$T = e^{-\omega^2 t}.$$

By the form of the Laplacian in spherical coordinates (11.15),

$$R'' + \frac{2}{\rho}R' = -\omega^2 R, \tag{11.61}$$

where the prime denotes differentiation with respect to ρ.

The product rule and some algebra bring (11.61) to the form

$$(\rho R)'' = -\omega^2 \rho R. \tag{11.62}$$

That is, $y = \rho R$ solves $y'' + \omega^2 y = 0$. Hence

$$\rho R = a \cos \omega \rho + b \sin \omega \rho,$$

i.e.,

$$R = a \frac{\cos \omega \rho}{\rho} + b \frac{\sin \omega \rho}{\rho}.$$

Insisting on bounded solutions means that $a = 0$. Imposing the zero surface condition (ii) yields

$$\sin \omega = 0,$$

i.e.,

$$\omega = n\pi, \quad n = 1, 2, \ldots .$$

Therefore, the temperatures within the cooling ball are

$$u(\rho, t) = \sum_1^\infty c_n e^{-n^2 \pi^2 t} \cdot \frac{\sin n\pi \rho}{\rho}, \qquad (11.63)$$

a sum of damped *sinc* functions. Initially,

$$1 = \sum_1^\infty c_n \frac{\sin n\pi \rho}{\rho}.$$

But the spatial eigenmodes $R_n = (\sin n\pi\rho)/\rho$ are mutually orthogonal. Hence (exercise)

$$c_n = \frac{2(-1)^{n+1}}{n\pi}.$$

Let us turn to several *steady-state problems*, where solutions are not changing with time.

Problem 7. (Steady state in a square)

$$\nabla^2 u = 0, \quad 0 < x, y < 1, \qquad (11.64)$$

subject to

(i) $u(0, y) = 0, \quad 0 \le y \le 1$

(ii) $u(1, y) = 1, \quad 0 < y < 1$

(iii) $u(x, 0) = 0, \quad 0 \leq x \leq 1$

(iv) $u(x, 1) = 0, \quad 0 \leq x \leq 1$

(see Figure 11.3).

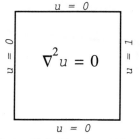

Figure 11.3. Steady problem.

Solution. Allow x to play the role of t in a transient problem:

$$\frac{\partial^2 u}{\partial x^2} = -\frac{\partial^2 u}{\partial y^2}.$$

We then proceed as before, finding special separated solutions to (i), (iii), and (iv), then combine these special solutions to satisfy the troublesome condition (ii).

Assume a solution

$$u(x, y) = X(x)Y(y)$$

and impose Laplace's equation to obtain

$$-\frac{d^2 X}{dx^2} / X = \frac{d^2 Y}{dy^2} / Y = \lambda = -\omega^2 \qquad (11.65)$$

and hence

$$X = a \cosh \omega x + b \sinh \omega x$$

and

$$Y = c \cos \omega y + d \sin \omega y.$$

Applying (iii) and (iv) to Y and i) to X yields

$$X_n = \sinh n\pi x$$

$$Y_n = \sin n\pi y,$$

resulting in the solution

$$u(x,y) = \sum_{n=1}^{\infty} c_n \sinh n\pi x \, \sin n\pi y. \tag{11.66}$$

Imposing the remaining condition (ii) gives

$$1 = \sum_{n=1}^{\infty} c_n \sinh n\pi \, \sin n\pi y, \tag{11.67}$$

so again aping Problem 1,

$$c_n = \begin{cases} \dfrac{4}{n\pi \sinh n\pi} & \text{if } n \text{ odd} \\[2ex] 0 & \text{if } n \text{ even.} \end{cases} \tag{11.68}$$

Problem 8. Solve the Poisson problem

$$\nabla^2 u = -1, \qquad 0 < x, y < 1, \tag{11.69}$$

subject to zero boundary conditions.

Solution. Taking the Rellich-guaranteed complete orthonormal basis consisting of the eigenfunctions of the Laplacian on the square with zero boundary conditions,

$$\phi_{mn}(x,y) = 2 \sin m\pi x \, \sin n\pi y, \tag{11.70}$$

belonging to the eigenvalue

$$\lambda_{mn} = -\pi^2(m^2 + n^2), \tag{11.71}$$

merely write down the solution

$$u(x,y) = \sum_{m,n=1}^{\infty} \frac{c_{mn}\phi_{mn}(x,y)}{\lambda_{mn}} \tag{11.72}$$

where the coefficients c_{mn} (Exercise 13.8) come from the expansion

$$-1 = \sum_{m,n=1}^{\infty} c_{mn}\phi_{mn}(x,y). \tag{11.73}$$

Remark 7. The solution form (11.72) is a prescription for solving many Poisson problems $\nabla^2 u = f(x)$ subject to Dirichlet boundary conditions on a bounded domain Ω. Throw the boundary conditions onto the right-hand side by taking any twice continuously differentiable function $h(x)$ on the domain Ω with boundary conditions that agree with u; subtracting h from u yields a Poisson problem $\nabla^2 v = g$ with zero boundary conditions. Find the complete orthonormal basis of eigenfunctions $\nabla^2 \phi_n = \lambda_n \phi_n$ of the Laplacian that is guaranteed by Rellich. Write the solution as

$$v(x) = \sum_{n=1}^{\infty} \frac{\langle \phi_n, g \rangle}{\lambda_n} \phi_n = \langle \kappa(x,y), g(y) \rangle = \int_{\Omega} \kappa(x,y) g(y) \, dy \quad (11.74)$$

where the *kernel* κ (Green's function) is the series

$$\kappa(x,y) = \sum_{n=1}^{\infty} \frac{\phi_n(x)\phi_n(y)}{\lambda_n}. \quad (11.75)$$

Note that (11.74) is at least formally a solution to $\nabla^2 v = g$ since

$$\nabla_x^2 v(x) = \nabla_x^2 \int_{\Omega} \kappa(x,y) g(y) \, dy$$

$$= \int_{\Omega} \nabla_x^2 \kappa(x,y) g(y) \, dy = \int_{\Omega} \sum_{n=1}^{\infty} \nabla_x^2 \frac{\phi_n(x)\phi_n(y)}{\lambda_n} g(y) \, dy$$

$$= \int_{\Omega} \sum_{n=1}^{\infty} \phi_n(x)\phi_n(y) g(y) \, dy = \sum_{n=1}^{\infty} \phi_n(x) \int_{\Omega} \phi_n(y) g(y) \, dy$$

$$= \sum_{n=1}^{\infty} \langle \phi_n, g \rangle \phi_n(x) = g(x).$$

Problem 9. (Quantum particle) A small particle is roaming somewhere within a potential-free box $0 < x, y, z < 1$. Find its likely location.

Solution. Whereas classical mechanics is a collection of relationships between measured quantities, quantum mechanics consists of the analogous relationships between the instruments taking the observations. The mathematical analog of an instrument is a *Hermetian* (self-adjoint) operator A where on average, the measurement yields the value $\langle A\psi, \psi \rangle$. The *wave function* ψ is the general solution (with $\|\psi\|_2 = 1$) to *Schrödinger's equation*,

$$i\hbar \frac{\partial \psi}{\partial t} = H\psi = T\psi + V\psi, \qquad (11.76)$$

where H, T, and V observe total, kinetic, and potential energy, respectively. For a single particle of mass m the instrument T has the form

$$T\psi = -\frac{\hbar^2}{2m}\nabla^2 \psi. \qquad (11.77)$$

The instrument V measuring potential energy is given by multiplication by the classical potential

$$V\psi = v(x)\psi. \qquad (11.78)$$

For our given problem, after going over to nondimensional variables, we must solve Schrödinger's equation,

$$i\frac{\partial \psi}{\partial t} = -\nabla^2 \psi + 0 \cdot \psi, \qquad (11.79)$$

subject to zero boundary conditions.

Assuming that $\psi = T(t)X(x)Y(y)Z(z)$, (11.79) becomes

$$-i\frac{\dot{T}(t)}{T(t)} = \frac{X''(x)}{X(x)} + \frac{Y''(y)}{Y(y)} + \frac{Z''(z)}{Z(z)} = -E, \qquad (11.80)$$

where E is the (positive) *eigenenergy*. By our now routine calculation,

$$X_k(x) = \sin k\pi x \qquad (11.81a)$$

$$Y_l(y) = \sin l\pi y \qquad (11.81b)$$

$$Z_m(z) = \sin m\pi z \qquad (11.81c)$$

yielding the *eigenstate* or *stationary state*

$$\psi_{klm} = 2\sqrt{2} \sin k\pi x \, \sin l\pi y \, \sin m\pi z \qquad (11.82a)$$

of norm 1 belonging to the energy level

$$E_{klm} = \pi^2(k^2 + l^2 + m^2). \qquad (11.82b)$$

Thus the wave function is

$$\psi = \sum_{k,l,m=1} c_{klm} e^{-iE_{klm}t}\psi_{klm} \tag{11.83}$$

with ψ normalized by

$$\sum_{k,l,m=1} |c_{klm}|^2 = 1.$$

§11.4 Unbounded Spatial Domains

Problem 10. Find the step response of an early undersea telegraph cable.

Solution 1. (Operationally) Early undersea telegraph cables suffered from *intersymbol distortion,* where transmitted pulses would smear spatially as they traveled down the cable, interfering with previous and subsequent pulses [Nahin; MacCluer, 1994]. Kelvin examined the problem and proposed a diffusion model for the voltage $u = u(x,t)$ at location x at time t: In nondimensional variables,

$$\frac{\partial u}{\partial t} = \frac{\partial^2 u}{\partial x^2}, \quad 0 < x, t < \infty, \tag{11.84}$$

subject to

(i) $u(x,0) = 0, \ 0 < x$

(ii) $u(0,t) = 1, \ t \geq 0$

(iii) $u(\infty,t) = 0, \ t \geq 0$

Taking the solution $u(x,t)$ over to the frequency domain as $U(x,s)$, (11.84) becomes

$$sU = U'', \tag{11.85}$$

so

$$U(x,s) = c(s)e^{-\sqrt{s}x} + d(s)e^{\sqrt{s}x}. \tag{11.86}$$

Imposing the right boundary condition at infinity, $U(\infty,s) = 0$, means that $d(s) = 0$. Imposing the left condition $U(0,s) = 1/s$ gives our solution,

$$U(x,s) = \frac{e^{-\sqrt{s}x}}{s}. \tag{11.87}$$

Consulting tables [Råde and Westergren], we have our sought-for time-domain solution

$$u(x,t) = \operatorname{erfc}\left(\frac{x}{2\sqrt{t}}\right), \tag{11.88}$$

where the *error function complement*

$$\operatorname{erfc}(x) = \frac{2}{\sqrt{\pi}} \int_x^\infty e^{-\beta^2}\, d\beta. \tag{11.89}$$

As long as we are here, note that since the impulse response is the generalized derivative of the step response,

$$s\left\{\operatorname{erfc}\left(\frac{x}{2\sqrt{t}}\right)\right\} = e^{-\sqrt{sx}} = \left\{\frac{x}{2\sqrt{\pi}t^{3/2}}e^{-x^2/4t}\right\} \tag{11.90}$$

(Exercise 11.24). Thus the response $u = u(x,t)$ of the cable at a distance x at time t to a arbitrary input signal $u(0,t) = f(t)$ at the left end is the convolution

$$u(x,t) = \frac{x}{2\sqrt{\pi}\,t^{3/2}}e^{-x^2/4t} * f(t). \tag{11.91}$$

Solution 2. (Separation of variables) Using the rule of thumb *subtract off the steady state*, we instead solve for

$$v(x,t) = 1 - u(x,t) = T(t)X(x) \tag{11.92}$$

to obtain

$$\frac{\dot{T}(t)}{T(t)} = \frac{X''(x)}{X(x)} = \lambda = -\alpha^2,$$

giving

$$T(t) = e^{-\alpha^2 t} \tag{11.93}$$

and

$$X(x) = a\cos\alpha x + b\sin\alpha x. \tag{11.94}$$

Imposing the left boundary condition on v, hence X, we see that

$$X(x) = \sin\alpha x.$$

Because the spatial domain is infinite, there are no additional conditions that would bind α to a discrete list. The problem has continuous spectrum. Instead, to achieve the initial condition we must superimpose the separated solutions $e^{-\alpha^2 t}\sin\alpha x$ continuously to form

$$v(x,t) = \int_0^\infty c(\alpha)e^{-\alpha^2 t}\sin\alpha x\, d\alpha \tag{11.95}$$

with a continuous weighting $c(\alpha)$. Imposing the initial conditions leads to

$$1 = \int_0^\infty c(\alpha) \sin \alpha x \, d\alpha. \tag{11.96}$$

But there is the famous *sinc formula*

$$\frac{2}{\pi} \int_0^\infty \frac{\sin b\alpha}{\alpha} d\alpha = \text{sgn}(b), \tag{11.97}$$

where the integral converges in the Cauchy principal value (PV) sense. Thus taking weight $c(\alpha) = 2/\pi\alpha$,

$$u(x,t) = 1 - v(x,t) = 1 - \frac{2}{\pi} \int_0^\infty e^{-\alpha^2 t} \frac{\sin \alpha x}{\alpha} d\alpha. \tag{11.98}$$

There are numerous delicate convergence problems present in solutions of this type (see [MacCluer, 1994]).

§11.5 Periodic Steady State

Let us now consider systems that are being driven by sinusoidal inputs well after initial startup transients have died away. Such problems yield to the *Helmholtz phasor method*.

Problem 11. How deeply within the Earth is felt the annual temperature cycle at the surface?

Solution. Assume that daily mean temperature follows a sinusoid about some yearly mean value. Let $u = u(x,t)$ be the earth temperature at depth x at time t referenced to this yearly mean temperature and normalized by the maximum variation. Then the model is $\partial u/\partial t = \nabla^2 u$ subject to the periodic condition $u(0,t) = \sin \omega t$ and $u(\infty, t) = 0$. It is intuitively clear that the solution must be of the form

$$u(x,t) = X(x) \sin[\omega t - \theta(x)], \tag{11.99}$$

where $X(x)$ is the decreasing amplitude of the effect of the driving surface temperature and where $\theta(x)$ is the phase lag at depth x. Placing this solution form into the heat equation and attempting to solve for X and θ would be a disaster. (Try it!)

Instead, let us employ a classical trick associated with the name Helmholtz: Search for a *phasor* solution

$$\rho(x,t) = e^{i\omega t - i\theta(x)} X(x) = e^{i\omega t} Y(x), \tag{11.100}$$

where our actual physical temperature u forms the imaginary part. Placing the phasor ρ into the heat equation and canceling the common factor $e^{i\omega t}$ reduces the problem to a single ODE

$$i\omega Y(x) = Y''(x) \tag{11.101}$$

with general solution

$$Y(x) = ce^{-\zeta x\sqrt{\omega}} + de^{\zeta x\sqrt{\omega}}, \tag{11.102}$$

where ζ is a square root of i, viz., $\zeta = (1+i)/\sqrt{2}$. Because temperatures are bounded, $d = 0$. Imposing the surface condition

$$\rho(0, t) = e^{i\omega t} \tag{11.103}$$

gives that $Y(0) = c = 1$. Thus

$$Y(x) = e^{-\zeta x\sqrt{\omega}} = e^{-i\theta(x)} X(x), \tag{11.104}$$

yielding the *attenuation* at depth x

$$X(x) = e^{-x\sqrt{\omega/2}} \tag{11.105a}$$

and *phase lag*

$$\theta(x) = x\sqrt{\omega/2}. \tag{11.105b}$$

Going back to dimensioned variables, these equations become

$$X(x) = e^{-x\sqrt{\omega/2\alpha}} \tag{11.106a}$$

and

$$\theta(x) = x\sqrt{\omega/2\alpha} \tag{11.106b}$$

where α is the diffusivity of the material.

The reciprocal quantity

$$\delta = \frac{\sqrt{2\alpha}}{\sqrt{\omega}} = \frac{\sqrt{\alpha}}{\sqrt{\pi f}} \tag{11.107}$$

is called the *depth of penetration,* the depth at which amplitude variations have dropped to $e^{-1} = 37\%$ of the surface amplitude.

For soil the depth of penetration is about 8 feet. In contrast, radio frequency at 100 MHz will penetrate only 0.0066 mm into a copper conductor. This rapid attenuation of surface variations at depths is called the *skin effect*. This skin effect limits radio communication with submarines to the extremely low frequency of 75 Hz and to depths of less than 300 feet; intersymbol distortion limits signal throughput to 1 bit per second.

§11.6 Other Distributed Models

Example. (Acoustics) The acoustic pressure $p = p(x, t)$ and particle (vector) velocity $V = V(x, t)$ within a spatial domain Ω are determined by the system

$$\frac{\partial p}{\partial t} = \rho c^2 \, \nabla \circ V \tag{11.108a}$$

$$\frac{\partial V}{\partial t} = (1/\rho)\nabla p, \tag{11.108b}$$

where ρ is the fluid density and where c is the speed of sound in this fluid. For example, the speed of sound c is 1087 feet/second in air at 32°F but 4922 feet/second in seawater.

Note that differentiating (11.108) through by t yields the wave equations

$$\frac{\partial^2 p}{\partial t^2} = c^2 \nabla^2 p \tag{11.109a}$$

$$\frac{\partial^2 V}{\partial t^2} = c^2 \nabla^2 V. \tag{11.109b}$$

Example. The four fundamental **equations of Maxwell** that describe all electromagnetic phenomena (in homogeneous isotropic linear stationary media) are

$$\nabla \circ D = \rho \tag{11.110a}$$

$$\nabla \times H = J + \frac{\partial D}{\partial t} \tag{11.110b}$$

$$\nabla \times E = -\frac{\partial B}{\partial t} \tag{11.110c}$$

$$\nabla \circ B = 0, \tag{11.110d}$$

together with the ancillary $D = \epsilon E$, $B = \mu H$, $J = \sigma E$ (see [Edminister]).

Example. (Transport) Consider the simplest of all PDEs:

$$\frac{\partial u}{\partial t} + c\frac{\partial u}{\partial x} = 0, \tag{11.111}$$

$-\infty < x, t < \infty$. Note that this equation possesses the right traveling-wave solution

$$u(x,t) = f(x - ct), \tag{11.112}$$

where f is determined by the initial condition $u(x,0) = f(x)$.

More generally, consider any

$$a(x,y)u_x + b(x,y)u_y = 0. \tag{11.113}$$

This can be read as the directional derivative in the direction (a,b) is 0; i.e., u is constant along a curve with tangent vector (a,b). Such curves have normal $(-b,a)$, so locally satisfy

$$-b\, dx + a\, dy = 0. \tag{11.114}$$

Solutions $\phi(x,y) = c$ of the differential form (11.114) are called the *characteristic* curves of the PDE (11.113), one per integration constant c. Then

$$u(x,y) = f(\phi(x,y)) \tag{11.115}$$

will at least locally solve (11.113) since

$$au_x + bu_y = af'(\phi)\phi_x + bf'(\phi)\phi_y = f'(\phi)(-ba + ba) = 0.$$

The solution $u = u(x,t)$ is constant along the characteristic curves.

Example. (Traffic Flow) Let $u = u(x,t)$ denote the density (concentration) of cars per unit mile of expressway at location x at time t. The integral

$$N(a,b,t) = \int_a^b u(x,t)\, dx \tag{11.116}$$

is the number of cars at time t found within the portion of the expressway between $x = a$ and $x = b$. Conservation dictates that

$$\int_a^b u_t(x,t)\, dx + F(b,t) - F(a,t) = 0, \tag{11.117}$$

where $F(x,t)$ is the *flow rate* of cars per unit time at time t past the location x. After dividing through by $\Delta x = b - a$ and going to the limit the conservation law becomes

$$\frac{\partial u}{\partial t} + \frac{\partial F}{\partial x} = 0, \tag{11.118}$$

a one-dimensional version of (11.128b) below. But

$$F(x,t) = v(x,t)u(x,t),$$

(11.119)

where $v(x,t)$ is the local velocity of traffic at location x at time t, giving the general model

$$u_t + (vu)_x = 0.$$

(11.120)

This is as far as mathematics can take us without real-world input.

It is observed that drivers slow as traffic density increases and with increasing traffic density ahead. So it is reasonable to assume that velocity $v = g(u, u_x)$, where $g(\alpha, \beta)$ is maximum at $(0,0)$ for nonnegative α and β. There are several accepted speed models. One is *Greenshields' linear model*

$$v = v_f\left(1 - \frac{u}{u_j}\right)$$

(11.121)

where v_f is the free flow speed with no traffic and where u_j is the jam density. There is *Greenberg's logarithmic model*,

$$v = v_m \ln(u_j/u)$$

(11.122)

where v_m is speed at maximum flow, and *Underwood's exponential model*

$$v = v_f e^{-u/u_m},$$

(11.123)

where u_m is density at maximum flow.

Let us assume as in the speed models above that the driver is not looking ahead but is responding only to local density, i.e., $v = g(u)$. Then the model (11.120) is of the form

$$u_t + a(u)u_x = 0,$$

(11.124)

where

$$a(u) = g'(u)u + g(u) = \frac{dg(u)u}{du} = \frac{dvu}{du} = \frac{dF}{du}.$$

(11.125)

Typically, the flow $F = vu = g(u)u$ plotted against density u appears as in Figure 11.4.

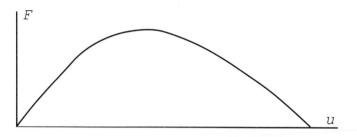

Figure 11.4. Traffic flow rate F versus density u .

Assume for the moment that traffic density $u = u(x,t)$ is continuously differentiable for $t \geq 0$. Solve the ODE

$$\frac{dx}{dt} = a(u(x,t)) \tag{11.126}$$

near (x_1, t_1), $t_1 \geq 0$ to obtain the characteristic curves $x = x(t)$ that
by Picard fill the half-plane $t \geq 0$ without meeting. The density u
must be constant on these characteristic curves since

$$\frac{d}{dt}u(x(t),t) = u_x \dot{x} + u_t = u_x a + u_t = 0.$$

But then since u is constant on these curves, $\dot{x} = a(u)$ must also be
constant; i.e. *the characteristic curves of* (11.124) *are straight lines,*
viz., $x - at = x_0$ where $a = a(u(x_0, 0))$, as in Figure 11.5.

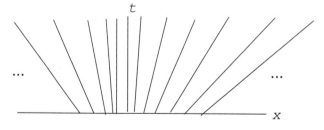

Figure 11.5. Characteristics of traffic density.

The slopes a of the characteristics $x - at = x_0$ must increase
with increasing x-intercept x_0, for otherwise characteristics will meet.
Moreover, traffic density is of the form

$$u(x,t) = f(x - at), \tag{11.127}$$

where initially $f(x) = u(x, 0)$.

But something is amiss — traffic is not this well behaved. Characteristics evidently must meet at certain times and places. Where they meet, singularities appear in the solution $u = u(x,t)$; characteristic lines meet at a *shock* that propagates in time and space.

For example, suppose that a slow-moving truck enters the expressway at $t = 0$ at location $x = x_0$, inducing a shock (the intersection of $x - a_0t = x_0$ with lines to its left) that propagates forward in time (up the line $x - a_0t = x_0$) but back down the expressway — a sudden jump in density, a compression, that moves in the negative x direction as in Figure 11.6. When the truck exits, a rarefaction will move up the expressway.

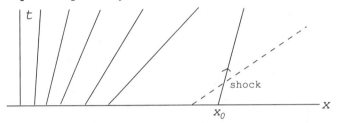

Figure 11.6. A slow-moving truck entering at x_0 induces a density shock that propagates forward in time but back down the freeway.

Such models are used to obtain a handle on rush hour congestion, to help in rational decision making. The literature on traffic modeling is extensive. A good place to start is the Transportation Research Board monograph by Gerlough and Huber, [May], and the bibliography found on pages 106–107 of *Mathematical Modelling* by Klamkin.

Example. The Navier–Stokes equations of fluid flow:

$$\frac{\partial V}{\partial t} + \frac{1}{\rho}\nabla p + (V \circ \nabla)V = \nu\nabla^2 V, \qquad (11.128a)$$

$$\rho_t + \nabla \circ \rho V = 0. \qquad (11.128b)$$

where ρ is the density, V the velocity, p pressure, and where $\nu = \mu/\rho$ is the kinematic viscosity of the fluid. The Navier–Stokes equations are the subject of *intense* study [McOwen, Strang].

There are more articles published yearly on PDEs than on the remainder of mathematics combined. The subject is central to mathematics, science, economics, and engineering. A beginner may start with [DuChateau and Zachmann], [Churchill and Brown], or [MacCluer, 1994]. Then consult the classic [Fourier], [Courant and Hilbert], [Sobolev], or [Ladyzhenskaya]. For an industrial flavor see [Friedman and Littman].

Exercises

11.1 Generically let f be a scalar field, F a vector field. Let $\nabla\cdot$, $\nabla\circ$, and $\nabla\times$ denote the gradient, divergence, and curl, respectively. Of the 18 symbols

$$
\begin{matrix}
\cdot & & \cdot & & f \\
\nabla & \circ & \nabla & \circ & \\
\times & & \times & & F
\end{matrix}
\quad,
$$

only five make sense. Which five? Of the five, two are identically 0. Which two?

11.2 Find a vector field F with no *potential*, i.e., where there is no f for which $F = -\nabla f$.

Hint: Give an infamous example such as 'dQ' $= c_V\,dT + nRT\,dV/V$ or '$\nabla\theta$' $= \nabla\arctan y/x$.

11.3 Find a sequence of change of variables that brings

$$
3\frac{\partial^2 u}{\partial x^2} + 2\frac{\partial^2 u}{\partial y^2} = 0
$$

to Laplace's equation (11.3).

11.4 Find a sequence of change of variables that brings

$$
\frac{\partial^2 u}{\partial x^2} + \frac{\partial^2 u}{\partial x\,\partial y} + \frac{\partial^2 u}{\partial y^2} = 0
$$

to Laplace's equation (11.3).

11.5 Find the change of variables that brings Schrödinger's equation for the electron of the hydrogen atom (in spherical coordinates)

$$
i\hbar\frac{\partial\psi}{\partial t} = -\frac{\hbar^2}{2m}\nabla^2\psi - \frac{e^2}{4\pi\epsilon_0}\frac{1}{\rho}\psi
$$

to nondimensional form

$$i\frac{\partial \psi}{\partial t} = -\nabla^2 \psi - \frac{1}{\rho}\psi.$$

11.6*Prove that the Laplacian $A = \nabla^2$ subject to zero boundary conditions on a bounded domain Ω is a *negative definite* operator, i.e., for $\phi \neq 0$ satisfying the boundary conditions,

$$\langle A\phi, \phi \rangle < 0.$$

Outline: Integrate once by parts via Green's first identity [O'Neil]:

$$\int_\Omega f\nabla^2 g \, dx = \int_{\partial\Omega} f\nabla g \circ \mathbf{n} \, ds - \int_\Omega \nabla f \circ \nabla g \, dx.$$

11.7 Prove that the eigenvalues of a negative definite operator are negative.

11.8*In fact, prove that the Laplacian $A = \nabla^2$ subject to zero boundary conditions on a bounded domain Ω is a *self-adjoint* operator, i.e.,

$$\langle A\phi, \psi \rangle = \langle \phi, A\psi \rangle$$

for all ϕ, ψ satisfying the boundary conditions.

Outline: Integrate twice by parts via Green's first identity.

11.9 Prove that eigenfunctions belonging to distinct eigenvalues of a self-adjoint operator are orthogonal.

Outline: If $A\phi = \lambda\phi$ and $A\psi = \mu\psi$, then $\lambda\langle\phi,\psi\rangle = \mu\langle\phi,\psi\rangle$.

11.10 Modify Problem 1 by replacing condition (iii) by $u(1,t) = 1$. Demonstrate that an attempt to separate variables now fails. Next, after applying the rule of thumb to *subtract off the steady state*, solve the modified problem by separation of variables.

11.11 Establish (11.26).

11.12 Solve Problem 1 modified to the nonhomogeneous $u_t = u_{xx} + 1$.

11.13 Find the first 20 eigenvalues $\lambda_n = -\omega_n^2$ of Problem 2 to six places.

Answer: $\lambda = -4.115858,\ -24.139342,\ -63.659107,\ \dots$.

Hint: Employ the Newton–Raphson algorithm to solve

$$\omega \cos \omega + \sin \omega = 0.$$

11.14 Establish (11.43) by repeatedly applying (11.40).

11.15 Establish Fourier's formulas (11.48).

11.16 Establish (11.51).

11.17 Show that the Laplacian of Problem 2 is self-adjoint and negative definite.

11.18 Make the substitution $x = \alpha r$ and $y = R(r)$ and reduce

$$\frac{d^2 R}{dr^2} + \frac{1}{r}\frac{dR}{dr} = -\alpha^2 R$$

to Bessel's equation

$$x^2 y'' + xy' + x^2 y = 0.$$

11.19 Find the fundamental frequencies of a (shear-free) cube of quivering Jello. (Model with the wave equation.)

11.20 Find how the (nondimensional) interior temperature of the square $-2 < x, y < 2$ with zero boundary temperature relaxes to 0 given that the initial temperature was 1 on the smaller square $-1 < x, y < 1$ and zero elsewhere.

11.21 Find the sound made by a square drum $0 \leq x, y \leq 1$.

Answer:

$$u(x, y, t) = \sum_{m,n=1}^{\infty} c_{mn} \cos \pi t \sqrt{m^2 + n^2}\ \sin m\pi x\ \sin n\pi y.$$

11.22 Argue from (11.76)–(11.78) that a quantum bouncing ball of mass m has wave function ψ satisfying

$$i\hbar \frac{\partial \psi}{\partial t} = -\frac{\hbar^2}{2m} \frac{\partial^2 \psi}{\partial x^2} + mgx\psi$$

within the domain $x > 0$. Find the nondimensional stationary states ψ where $-\psi'' + x\psi = E\psi$.

Answer: $\psi(x) = \mathrm{Ai}(x - E)$, where $\mathrm{Ai}(x)$ is the *Airy* function.

11.23 Find the nondimensional form of the quantum model of a mass–spring system (with classical model $m\ddot{x} + kx = 0$).

11.24 Establish (11.90).

11.25 Make your best guess of future shapes of the plucked string of Problem 4 as determined in (11.50) and (11.51). Verify your guess by solving the problem via finite divided differences

$$\frac{u_i^{j+1} - 2u_i^j + u_i^{j-1}}{\Delta t^2} = \frac{u_{i+1}^j - 2u_i^j + u_{i-1}^j}{\Delta x^2}$$

or by truncating the series solution (11.50) and graphing. You will be astonished.

11.26 (**Project**) In 1966 Mark Kac asked, "Can one hear the shape of a drum?" Must two drums of different shapes have different eigenvalues? Argue that this is an important question. (This question was resolved negatively in 1992 by Gordon, Webb, and Wolpert.) Illustrate its importance by outlining an application to petroleum exploration, antisubmarine warfare, the curing of composite materials, or the like.

11.27* As a rule of thumb, the natural vibrational frequencies of one (spatial)-dimensional bodies are more or less equally spaced, as in Problem 4. In contrast, two- or more dimensional bodies have vibrational frequencies that become more densely packed in the higher octaves. Illustrate this by plotting against f the number $N = N(f)$ of natural vibrational frequencies less than f of the square drum of Exercise 11.21. What is the asymptotic value of N?

Hint: The number of integral *lattice* points (m, n) within the circle of radius r is approximately the area of the circle.

Argue that active sound suppression within heating ducts is a far easier problem than sound suppression within an automobile or airplane fuselage.

11.28 Solve, by separating variables, the steady problem $\nabla^2 u = 0$ on the quarter plane $0 < x, y < \infty$ subject to $u(0, y) = 1$, $u(x, 0) = 0$, and u bounded.

Answer:

$$u(x, y) = \frac{2}{\pi} \int_0^\infty e^{-\alpha x} \frac{\sin \alpha y}{\alpha} d\alpha.$$

11.29 Separate variables to solve the steady problem $\nabla^2 u = 0$ on the doubly infinite solid rod $u = u(r, z)$, $r < 1$, $-\infty < z < \infty$ in polar coordinates with boundary conditions $u(1, z) = \text{sgn}(z)$.

Answer:

$$u(r, z) = \frac{2}{\pi} \int_0^\infty \frac{I_0(\alpha r)}{I_0(\alpha)} \frac{\sin \alpha z}{\alpha} d\alpha,$$

where the *modified Bessel function*

$$I_0(x) = J_0(ix) = \sum_{k=0}^\infty \frac{1}{k!k!} \left(\frac{x}{2}\right)^{2k}$$

satisfies the *modified Bessel equation*

$$x^2 y'' + xy' - x^2 y = 0.$$

11.30 In contrast to Exercise 11.29, the half-infinite rod steady problem $\nabla^2 u = 0$ for $r < 1$ and $0 < z < \infty$, subject to $u(1, z) = 0$, $u(r, 0) = 1$, and u bounded, has discrete spectrum with solution

$$u(r, z) = 2 \sum_{k=1}^\infty \frac{J_0(\alpha_k r)}{\alpha_k J_1(\alpha_k)} e^{-\alpha_k z}.$$

Can $v(x, t) = 1 - u(x, t)$ be extended oddly to solve the doubly infinite rod problem of Exercise 11.29?

11.31 Find the periodic steady-state temperature $u(x, t)$ of a homogeneous isotropic rod $u_t = u_{xx}$ subject to $u(0, t) = \sin t$ and $u(1, t) = 0$. Solve this problem three ways: by separating variables to obtain the full transient solution, operationally, and via Helmholtz phasors.

11.32 Find the periodic steady state of the transverse vibrations $u(x,t)$ of a beam $u_{tt} = -u_{xxxx}$ subject to the clamped conditions $u(0,t) = \sin t,\ u_x(0,t) = 0 = u(1,t) = u_x(1,t)$.

11.33 Find the lowest resonant frequency of a 10-foot-long organ pipe. (Take the speed of sound to be 1100 feet/second.)

Answer: 27.5 Hz.

11.34 Deduce from (11.128) that the velocity profile of steady flow of a viscous incompressible fluid down a square tube is modeled by Poisson's equation $\nabla^2 V = -1$ (nondimensional) with zero boundary conditions.

11.35 Solve $u_t + x u_x = 0$ via the method of characteristics.

11.36 Using the MATLAB command `bessel(n,x)`, superimpose the graphs of $J_n(x)$ on the interval $0 < x < 20$ for $n = 0, 1, 2, 3, 4$. What do you observe about the zeros of these functions?

11.37 (**Project**) Let $u = u(x,t)$ be the population density of a country, i.e.,

$$P(b,t) - P(a,t) = \int_a^b u(x,t)\,dx$$

is the population at time t between the ages of a and b. Let $q(x)$ be the probability of death less net immigration at age x per year. Argue that

$$u_t + u_x + qu = 0.$$

Solve analytically by multiplying through by the integrating factor $\exp[\int_0^x q(y)\,dy]$. Argue that the left boundary condition is of the form

$$u(0,t) = \int_0^\infty b(x)u(x,t)\,dx.$$

Simulate this model via the numerical model $u_{i+1}^{j+1} = u_i^j - h q_i u_i^j$ and experiment with various initial populations, birth, and death rates. Is this model reasonable?

11.38 (**Project**) Research and discuss the several accepted PDE models for the time evolution of demographics. Start with [Keyfitz and Flieger] and [Keyfitz].

11.39*Model a heavy truck as a moving point load $F(x,t) = \delta(x - vt)$ as it crosses a span. What natural frequencies of the span are excited?

11.40*(Project) Model, solve, and simulate a person jumping up and down on a portable grandstand plank.

11.41 (Project) The pueblo-dwelling native Americans of the southwest built homes with adobe walls of a critical thickness — the walls delayed the sun's flux until it was needed to heat the interior during the cold nights. Model this wall. Using actual material properties, choose the optimal thickness for a 24-hour repetitive cycle. Compare your results against actual traditional construction.

11.42 Show that any solution u of the one-dimensional wave equation $u_{tt} - c^2 u_{xx} = 0$ is the sum of two traveling waves $u(x,t) = f(x - ct) + g(x + ct)$. The functions f and g are determined within constants that sum to 0.

Hint: Set $\alpha = x - ct$ and $\beta = x + ct$, then show that $u_{\alpha\beta} = 0$.

11.43 In a recent tort action an overweight man sued his brother-in-law for reckless endangerment after falling (over land) while sliding down a cable strung from a tree to the opposite shore of his in-law's lake. The pudgy plantiff claimed that "the strap was jerked violently from my hands soon after I grabbed it and began to slide." Explain this violent jerk.

11.44 Solve $u_{tt} = c^2 u_{xx}$ subject to the initial conditions $u(x,0) = e^{-x}$ and $u_t(x,0) = \cos x$ using the method of characteristics of Exercise 11.42.

11.45 Expand the function $f(t) = -1$ if $t < 0$ and $f(t) = t$ if $t \geq 0$ in Fourier series on the interval $[-\pi, \pi]$.

11.46 Show that there are traveling-wave solutions $u(x,t) = \sin(\omega t - \kappa x)$ to the beam equation $u_{tt} + c^4 u_{xxxx} = 0$. But in contrast to the wave equation, these waves *disperse;* their speed $\nu = c\sqrt{\omega}$ depends on frequency — the higher frequencies outpace lower. Complex waveforms are not preserved. Simulate a transverse displacement waveform $u(x,t)$ with initial shape $u(x,0) = \sum c_n \sin nx$ as it travels down a steel beam.

11.47 Expand $f(t) = \operatorname{sgn} t$ in a Fourier series on the interval $[-1, 1]$.

11.48 Show that indeed for u given by (11.98), $u(\infty, t) = 0$ for $t > 0$.

Hint: Set $\beta = \alpha x$.

11.49 (**Project**) Propose a simple nondestructive experiment to determine the thermal diffusivity $\alpha = \kappa/\rho c$ of a sample of concrete. Perform the experiment.

11.50 (**Project**) Obtain a 4-foot length of plastic pipe with end cap. Predict its acoustic resonant frequency as in Exercise 11.33. Acoustically excite the tube and measure its resonant frequency using a musical instrument, or better, with a sound card, as suggested in Exercise 4.7.

11.51 Find the steady-state temperature within a half cylinder $0 \le r \le 1$, $0 \le \theta \le \pi$, $0 \le z \le 1$ where the flat faces are held at temperature 0 and the curved side at temperature 1.

Answer:

$$\frac{u}{16} = \sum_{m,n=1,\,odd}^{\infty} \frac{I_m(n\pi r)}{I_m(n\pi)} \frac{\sin m\theta}{m\pi} \frac{\sin n\pi z}{n\pi}.$$

11.52 Find the frequency-domain solution of

$$\frac{\partial u}{\partial t} = \frac{\partial^2 u}{\partial x^2}, \quad 0 < x < 1$$

subject to $u(x, 0) = 0$, $u(0, t) = 0$, $u(1, t) = f(t)$.

Answer:

$$U = \frac{\sinh x\sqrt{s}}{\sinh \sqrt{s}} F(s).$$

11.53 Find the impulse response (Green's function) $p(t)$ of the Exercise 11.52, i.e., the function

$$\{p(t)\} = \frac{\sinh x\sqrt{s}}{\sinh \sqrt{s}},$$

via separation of variables as follows: Solve for the step response (i.e., $f(t) \equiv 1$), then operate by s.

Answer:

$$p(t) = 2\pi \sum_{n=1}^{\infty} (-1)^{n+1} e^{-n^2\pi^2 t} n \sin n\pi x.$$

11.54 Find the potential u within the cylindrical capacitor $r < 1$, $0 < z < 1$ when the end at $z = 0$ and the curved side is at voltage 0, while the other end at $z = 1$ is at voltage 1.

Hint: The potential satisfies Laplace's equation (11.3).

Answer:

$$u = 2 \sum_{n=1}^{\infty} \frac{J_0(\alpha_n r)}{\alpha_n J_1(\alpha_n)} \frac{\sinh \alpha_n z}{\sinh \alpha_n}$$

where the α_n are the positive zeros of J_0.

11.55 The universally accepted model for vibrations of an isotropic homogeneous elastic solid is

$$\rho \frac{\partial^2 D}{\partial t^2} = (\lambda + \mu)\nabla \, \nabla \circ D + \mu \nabla^2 D$$

where $D = u\mathbf{i} + v\mathbf{j} + w\mathbf{k}$ is particle displacement, and where λ and μ are *Lamé's* constants.

Show by simple manipulation that

$$\rho \frac{\partial^2 (\nabla \circ D)}{\partial t^2} = (\lambda + 2\mu)\nabla^2 (\nabla \circ D)$$

and

$$\rho \frac{\partial^2 (\nabla \times D)}{\partial t^2} = \mu \nabla^2 (\nabla \times D).$$

Thus in the language of seismology, the *pressure* wave $p = \nabla \circ D$ from a distant shock — traveling with velocity $\sqrt{(\lambda + 2\mu)/\rho}$ — arrives first, followed later by the *shear* wave $S = \nabla \times D$ traveling at the lower velocity $\sqrt{\mu/\rho}$. Bringing up the rear in this impulse train is the destructive surface-traveling *longwave* allowed by the very complex boundary conditions at the shear-free surface [Kolsky].

11.56 Find the wave function of a particle trapped within a potential-free sphere.

Chapter 12
Divided Differences

Differential equations are modeled and solved numerically
as linear algebra problems.

Derivatives are replaced by difference quotients to obtain numerical models for ODEs and PDEs. We practice this approach on a first-order ODE and the predator–prey system. The "pepperoni" heat transfer problem is solved by an explicit method and by the implicit Crank–Nicolson method. The recommended Runge–Kutta method is displayed.

§12.1 Euler's Method

The simplest of all numerical methods for solving differential equations is *Euler's method*, where derivatives are replaced by difference quotients:

$$\frac{dy}{dx} \rightarrow \frac{y(x+h) - y(x)}{h}.$$

With this replacement the differential equation

$$y' = f(x, y), \text{ subject to } y(x_0) = y_0, \tag{12.1}$$

is solved numerically (approximately) by the routine

> *with small steps of size h,*
> *beginning at x_0 and y_0,*
> *do*
> $$y = y + h \cdot f(x, y)$$
> $$x = x + h$$
> *until the required x is reached*

231

Example. Let us solve the ODE

$$y' = 2xy \tag{12.2}$$

subject to the initial condition $y(0) = 1$ for $0 \le x \le 1$. The exact
answer by separation of variables is

$$y = e^{x^2}. \tag{12.3}$$

To proceed numerically we first augment MATLAB's function list by
saving the function script

```
function z = fn(x,y)
z = 2*x*y;
```
as the file fn.m. Next we invoke Routine 12.1.

```
% Routine 12.1  Euler's method
x=0;                     % initial x
y = 1;                   % initial y
h = 0.1;                 % step size
N = 11;                  % number of steps
                         %
for i = 1:N              % begin loop
    a(i) = x;            % store old x
    b(i) = y;            % store old y
    y = y + h*fn(x,y);   % next y
    x = x + h;           % next x
    end;                 % end loop
plot(a,b)                % plot solution
```

Running the script above yields a plot of our numerical solution
to (12.2) for $0 \le x \le 1$. Let us check this numerical solver against
the actual solution (12.3) by continuing with the script:

```
hold                     % holds previous plot
x = 0:0.01:1             % x values
y = exp(x.*x);           % actual solution (12.3)
plot(x,y,'.')            % plot dots
```

to obtain Figure 12.1. The figure reveals an increasing divergence of
the numerical solution from the actual solution $y = \exp(x^2)$. This
poor fit stems from two causes — the use of a rather large step size h,

and the fundamental inaccuracy of Euler's method. Euler's method is an $O(h^2)$ method, where error at the first step is in proportion to the square of the step size h [Strang] (Exercise 12.12). At each step, our next estimate is obtained along the tangent line to the solution through the previous point (x, y).

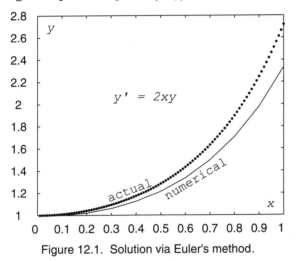

Figure 12.1. Solution via Euler's method.

Some improvement is possible via the *improved Euler* or *Heun method:*

> *with small steps of size h,*
> *beginning at x_0 and y_0,*
> *do*
> $$m_1 = f(x, y)$$
> $$m_2 = f(x + h, y + h \cdot m_1)$$
> $$y = y + h \cdot (m_1 + m_2)/2$$
> $$x = x + h$$
> *until the required x is reached*

Thus the present solution slope and the extrapolated slope are averaged to obtain the next estimate. In Exercise 12.1 you are asked to modify Routine 12.1 to an improved Euler method and compare its accuracy with the Euler's method solution of Figure 12.1. The improved Euler method is a (local) $O(h^3)$ method.

§12.2 Systems

Euler's method can be applied to systems without change using vector notation.

Example. (Predator–Prey) Consider the *Lotka–Volterra* equations:

$$\dot{x} = ax - bxy \tag{12.4a}$$

$$\dot{y} = -cy + dxy. \tag{12.4b}$$

Take the values $a = 1$, $b = 0.5$, $c = 0.75$, $d = 0.25$. Define the system with the function script

```
function zdot = fn(z)
zdot = [0,0]';
zdot(1) = z(1) - 0.5*z(1)*z(2);
zdot(2) = -0.75*z(2) + 0.25*z(1)*z(2); ,
```

save as **fn.m**, and run the script

```
% Routine 12.2   Euler's method on systems
t = 0;                  % start at t=0
z = [2,2]';             % initial population x,y
h = 0.005;              % step size
N = 1500;               % number of steps
                        %
for i = 1:N             % begin loop
   x(i) = z(1);         % save x
   y(i) = z(2);         % save y
   z = z + h*fn(z);     % next z
   t = t + h;           % (unnecessary in this problem)
   end;                 % end loop
plot(x,y)               % state space plot
```

to obtain the population trajectories of Figure 12.2. The trajectories are guaranteed to close by Exercise 9.33 given infinite precision.

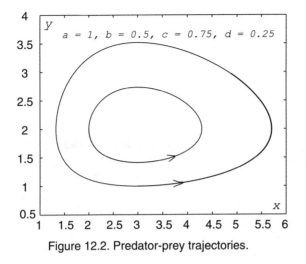

Figure 12.2. Predator-prey trajectories.

§12.3 PDEs

An extension of Euler's method can also numerically solve PDEs.

Problem. Suppose that a long pepperoni sausage is initially at room temperature when suddenly one end is placed against a hot source while the other end is held at room temperature. Find the evolution of the temperature regime within the sausage as it rises to its final steady state.

Solution 1. (Euler's method) After going to nondimensional variables the problem becomes

$$u_t = u_{xx}, \qquad 0 < x < 1, \qquad\qquad (12.5)$$

subject to $u(x,0) = 0$, $u(0,t) = 0$, $u(1,t) = 1$. We replace the partial derivatives with divided differences:

$$u_t(x,t) \rightarrow \frac{u(x,t+\Delta t) - u(x,t)}{\Delta t} \qquad\qquad (12.6a)$$

and

$$u_{xx}(x,t) \rightarrow \frac{\frac{u(x+\Delta x)-u(x,t)}{\Delta x} - \frac{u(x,t)-u(x-\Delta x,t)}{\Delta x}}{\Delta x} \qquad (12.6b)$$

to obtain the difference equation

$$u(x,t+\Delta t) = u(x,t) + h[u(x+\Delta x) - 2u(x,t) + u(x-\Delta x,t)], \quad (12.7)$$

where $h = \Delta t/\Delta x^2$. To implement this algorithm, divide the spatial domain $[0, 1]$ into n equal subintervals of length $\Delta x = 1/n$, declare the temperature on the ith subinterval during the jth interval of time Δt to be u_i^j, and set $h = \Delta t/\Delta x^2$ to obtain the recursion

$$u_i^{j+1} = u_i^j + h[u_{i+1}^j - 2u_i^j + u_{i-1}^j], \qquad (12.8)$$

$i = 1, 2, \ldots, n$ and $j = 0, 1, 2, 3, \ldots$. Impose the boundary conditions with fictitious cells $u_0^j = 0$ and $u_{n+1}^j = 1$ for all j. Impose the zero initial condition with the requirement that $u_i^0 = 0$ for $i = 1, 2, \ldots, n$.

Copy the formula (12.8) into the $(j + 1)$th row or column (depending on your recalculation pattern) of a spreadsheet and initiate recalculation [Orvis].

But the recursion (12.8) is actually a linear algebra problem:

$$\begin{vmatrix} 0 \\ u_1^{j+1} \\ u_2^{j+1} \\ \cdot \\ \cdot \\ \cdot \\ u_n^{j+1} \\ 1 \end{vmatrix} = \begin{pmatrix} 1 & 0 & 0 & . & . & . & 0 \\ h & 1-2h & h & 0 & . & . & . \\ 0 & h & 1-2h & h & 0 & . & . \\ . & & . & & & . & \\ & & & . & & & \\ 0 & . & & . & h & 1-2h & h \\ 0 & 0 & . & . & . & . & 1 \end{pmatrix} \begin{vmatrix} 0 \\ u_1^j \\ u_2^j \\ \cdot \\ \cdot \\ \cdot \\ u_n^j \\ 1 \end{vmatrix} ; \quad (12.9)$$

i.e., the interative process (12.8) is the repeated application of a matrix operator A,

$$u^{j+1} = Au^j \qquad (12.10)$$

beginning with the initial vector $u^0 = (0, 0, 0, \ldots, 0, 1)^T$. If all is well, this time-indexed sequence of approximate solutions

$$u^j = A^j u^0 \qquad (12.11)$$

will converge to the steady-state solution u^∞, a fixed point of the operator A.

So it is with no great surprise that MATLAB is effective on this and similar problems.

```
% Routine 12.3  Heat equation via Euler
n = 50;                    % number of spatial cells
J = 1000;                  % number of time steps
h = 0.4;                   % step size h = dt/dx*dx
```

```
                              % build the matrix H as follows:
Id = eye(n+2);                % identity matrix
H = ones(n+2);                % matrix of all 1's
H = tril(H,1);                % lower diag.
H = triu(H,-1);               % tridiag.  of 1's
H = H - 3*Id;                 % -2's on diag.
H(1,1)=0;                     % corrections to H:
H(1,2) = 0;                   %
H(n+2,n+1) = 0;               %
H(n+2,n+2) = 0;               %
A = Id + h*H;                 % form A
C = A';                       %
                              %
u = zeros(1,n+2);             % size u
u = 0*u;                      % impose initial conditions
u(0) = 0;                     % impose left boundary conds.
u(n+2) = 1;                   % impose right boundary conds.
                              % iterate:
for j=1:J                     %
   u = [u;u(j,:)*C];          % stack up temps.  as rows
   end;                       %
                              % plot selected snapshots
```

Plotting every 100th iteration yields the snapshots of Figure 12.3. Notice how successive iterations are approaching the intuitively clear steady state $u = x$.

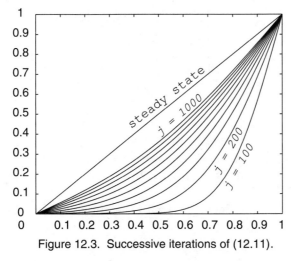

Figure 12.3. Successive iterations of (12.11).

Alert. There are two issues present in any linear iteration

$$u^j = Au^{j-1} = A^j u^0$$

that serves as a consistent numerical model of an ODE or PDE problem: Is the process *stable*, and if so, is it *convergent* to the actual solution. The simple *explicit* algorithm (12.8) becomes unstable for $h \geq 1/2$, as can be seen experimentally (Exercise 12.6) and analytically (Exercise 12.8). Thus you are not permitted to take large time steps Δt or to use small cell sizes Δx that would result in $h = \Delta t / \Delta x^2 \geq 0.5$. The instability can be remedied by going over to an *implicit* algorithm such as *Crank–Nicolson*.

Solution 2. (Crank–Nicolson) Write the operator A of (12.9) as

$$A = \begin{pmatrix} 1 & 0 & 0 & . & . & . & 0 \\ h & 1-2h & h & 0 & . & . & . \\ 0 & h & 1-2h & h & 0 & . & . \\ & . & & . & & & \\ & & & . & & & \\ 0 & . & & . & h & 1-2h & h \\ 0 & 0 & & . & . & . & 1 \end{pmatrix}$$

$$= \begin{pmatrix} 1 & 0 & 0 & . & . & . & 0 \\ 0 & 1 & 0 & 0 & . & . & . \\ 0 & 0 & 1 & 0 & 0 & . & . \\ & . & & . & & & \\ & & & . & & & \\ 0 & . & . & . & 0 & 1 & 0 \\ 0 & 0 & . & . & . & . & 1 \end{pmatrix} + h \begin{pmatrix} 0 & 0 & 0 & . & . & . & 0 \\ 1 & -2 & 1 & 0 & . & . & . \\ 0 & 1 & -2 & 1 & 0 & . & . \\ & . & & . & & & \\ & & & . & & & \\ 0 & . & & . & 1 & -2 & 1 \\ 0 & 0 & & . & . & . & 0 \end{pmatrix}$$

$$= I + hH \qquad\qquad (12.12)$$

In this notation the model (12.8) becomes

$$u^{j+1} = (I + hH)u^j. \qquad\qquad (12.13)$$

Replace this numerical model (12.13) with the implicit *Crank–Nicolson* model

$$u^{j+1} = u^j + \frac{h}{2}(Hu^j + Hu^{j+1}), \qquad\qquad (12.14)$$

i.e.,

$$\left(I - \frac{h}{2}H\right)u^{j+1} = \left(I + \frac{h}{2}H\right)u^{j}, \tag{12.15}$$

giving the explicit solution

$$u^{j+1} = \left(I - \frac{h}{2}H\right)^{-1}\left(I + \frac{h}{2}H\right)u^{j}. \tag{12.16}$$

The good news is that the required inversion of $I-(h/2)H$ to exploit the unconditionally stable algorithm (12.16) is easy to code: Starting at the upper left-hand corner, recursively bring the tridiagonal $I - hH/2$ to upper bidiagonal form, then back substitute. Of course, MATLAB does the additional work for you with one line of code.

```
% Routine 12.4  Heat equation via Crank-Nicolson
n = 50;                 % number of spatial cells
J = 1000;               % number of time steps
h = 0.4;                % step size h = dt/dx*dx
                        % build the matrix H as follows:
Id = eye(n+2);          % identity matrix
H = ones(n+2);          % matrix of all 1's
H = tril(H,1);          % lower diag.
H = triu(H,-1);         % tridiag.  of 1's
H = H - 3*d;            % -2's on diag.
H(1,1)=0;               % corrections to H:
H(1,2) = 0;             %
H(n+2,n+1) = 0;         %
H(n+2,n+2) = 0;         %
A = Id + h*H;           % form iterator A:
A = inv(Id - h*H)*A;    % Crank-Nicolson!
C = A';                 %
                        %
u = zeros(1,n+2);       % size u
u = 0*u;                % impose initial conditions
u(1) = 0;               % impose left boundry conds.
u(n+2) = 1;             % impose right boundary conds.
                        % do the iterations:
for j=1:J               %
   u = [u;u(j,:)*C];    % stack up temps.  as rows
   end;                 %
                        % plot selected snapshots
```

In Exercise 12.9 you are encouraged to experiment with Crank-Nicolson (Routine 12.4), to explore its accuracy and stability. To read further about stability issues, browse the wonderful *Introduction to Applied Mathematics* by Strang.

§12.4 Runge–Kutta Method

The recommended algorithm for problems

$$y' = f(x, y) \tag{12.17}$$

subject to $y(x_0) = y_0$, provided that f is not costly to evaluate, is the *Runge–Kutta* fourth-order method:

$$y_{\text{next}} = y + \frac{h}{6}(k_1 + 2k_2 + 2k_3 + k_4) \tag{12.18a}$$

where

$$
\begin{aligned}
k_1 &= f(x, y), \\
k_2 &= f(x + h/2, y + hk_1/2), \\
k_3 &= f(x + h/2, y + hk_2/2), \\
k_4 &= f(x + h, y + hk_3).
\end{aligned}
\tag{4.18b}
$$

The local error for this method is a respectable $O(h^5)$; i.e., if step size h is halved, error at the first step will decrease to $1/32$ of the previous error. MATLAB provides this routine as **ode45**.

Example. Let us employ MATLAB's built-in solver to again solve the Lotka–Volterra equations (12.4). Run the following script to again obtain Figure 12.2.

```
% Routine 12.5  Runge-Kutta method
T = [0,7];                   % set time span
z0 = [2,2]';                 % initial populations x,y
 [t,z] = ode45('fn',T,z0);   % do Runge-Kutta
 x = z(:,1);                 % x coordinate
 y = z(:,2);                 % y coordinate
plot(x,y)                    % plot the state space orbit
```

Exercises

12.1 Modify Routine 12.1 to an improved Euler method and compare its accuracy with Euler's method on problem (12.2) graphically and in tabular form.

12.2 Experiment with Routine 12.1 by varying step size h. Is there a point where decreasing h does not yield increased accuracy?

12.3 Experiment with Routine 12.2 by choosing various initial population sizes (x, y). Avoid the equilibrium population $(c/d, a/b)$.

12.4 Use Euler's method to investigate several trajectories of the very interesting system (from [Boyce and diPrima]):

$$\dot{x} = x + y - x(x^2 + y^2)$$
$$\dot{y} = -x + y - y(x^2 + y^2).$$

Look at trajectories that initiate within and without the unit circle $x^2 + y^2 = 1$.

12.5 Solve (12.5) by separating variables after subtracting off the steady state. Test Routine 12.1 against the exact solution. Compare the benefits (if any) of small step size h and small cell size $\Delta x = 1/n$.

12.6 Using Routine 12.3 verify experimentally that the Euler routine (12.8) is unstable for $h \geq 0.5$.

12.7* Show that an iteration scheme $u^j = A^j u^0$ converges for every initial point u^0 if and only if the eigenvalues λ of A distinct from 1 have modulus $|\lambda| < 1$ *and* the eigenspace belonging to $\lambda = 1$ is full rank, i.e., its rank equals the multiplicity of the root $\lambda = 1$ in the characteristic equation $\phi(\lambda) = \det(\lambda I - A)$.

Hint: Reduce to the case where A has only one distinct eigenvalue.

12.8* Show analytically that the Euler algorithm (12.8) is stable if and only if $h < 0.5$ by showing that H of (12.12) has eigenvalues between -4 and 0 (see Exercise 12.13).

12.9 Experiment with the Crank–Nicolson Routine 12.4. Can you destablize the routine by increasing h? Compare its accuracy against the Euler Routine 12.3 for equal h.

12.10 Experiment with Routine 12.3 or 12.4 by changing initial and boundary conditions.

12.11 After going over to phase space, use Routine 12.2 to numerically solve $\ddot{x} + c\dot{x} + 3x = \sin t$. Plot trajectories for various damping coefficients c.

12.12* Assuming that the second partial derivatives of $f(x, y)$ are continuous and bounded on a region, show that Euler's method and the improved Euler method within this region have at the first step a *local truncation error* of $O(h^2)$ and $O(h^3)$, respectively, from the actual solution.

Hint: $y(x + h) = y(x) + y'(x)h + y''(c)h^2/2$ and $f(x, y(x))' = f_x(x, y(x)) + f_y(x, y(x))y'(x)$.

12.13 Prove *Gerschgorin's theorem:* Every eigenvalue λ of the $n \times n$ matrix $A = (a_{ij})$ lies in at least one of the disks

$$|\lambda - a_{ii}| \leq r_i = \sum_{j \neq i}^{n} |a_{ij}|.$$

Hint: Look at the relation $Ax = \lambda x$ row by row.

12.14 Experimentally investigate the nature of the solutions of *van der Pol's equation* in state space: $\dot{x} = y$, $\dot{y} = -x + \epsilon(1 - x^2)y$ for $\epsilon \geq 0$.

12.15 Write your own Runge–Kutta routine. Tabulate your results against the results of Routine 12.1.

12.16 Plot various trajectories in phase space (q, p) of damped harmonic motion (9.28) using MATLAB's solver ode45.

12.17 (**Project**) Solve the *three-body problem:* Model, code, and simulate the motions of a sun and two massive interacting planets. Do the special case where the sun is essentially stationary and motion is planar. Assume all three masses to be point masses. Will collisions occur? Will a planet spiral in to be consumed by the sun? Can a planet escape the system? Plot selected motions.

Chapter 13

Galerkin's Method

Galerkin's method yields astonishing accuracy with small effort.

We learn how to apply Galerkin's numerical method to ODEs, to eigenvalue problems, to steady and transient PDE problems, and to practice the art of finite elements. We conclude with an explanation of why this method is so effective.

§13.1 Galerkin's Requirement

Our objective is to numerically solve a linear differential equation $L(u) = f$ by choosing *basis* (*trial*) functions ϕ_j, then approximating the actual solution u by a linear combination of these functions:

$$\tilde{u} = \sum_{j=1}^{N} c_j \phi_j. \tag{13.1}$$

Among all choices for the coefficients c_j, we insist that the approximation $\tilde{u}$ satisfies

Galerkin's Requirement. The *residual* $R = L(\tilde{u}) - f$ must be orthogonal to the basis elements $\phi_1, \phi_2, \ldots, \phi_N$ used in the approximation, i.e.,

$$\langle \phi_i, R \rangle = \sum_{j=1}^{N} c_j \langle \phi_i, L\phi_j \rangle - \langle \phi_i, f \rangle = 0, \tag{13.2}$$

a system of N equations in the N unknowns c_j. The basis functions must be independent, satisfy the homogeneous boundary conditions of the problem $Lu = f$, and be dense in a space containing the actual solution.

Example 1. Let us numerically solve $y' + y = 1$ on $0 < x < 1$ subject to $y(0) = 0$. The actual solution is (exercise) $y(x) = 1 - e^{-x} = x - x^2/2! - x^3/3! + \cdots$. Use as basis functions the powers x, x^2, x^3, $\ldots$, each of which satisfies the boundary condition $y(0) = 0$. Then the trial solution

$$\tilde{y} = \sum_{j=1}^{N} c_j x^j \tag{13.3}$$

to our problem $Ly - 1 = y' + y - 1 = 0$ yields the residual

$$R = L\tilde{y} - 1 = -1 + \sum_{j=1}^{N} c_j (j x^{j-1} + x^j). \tag{13.4}$$

Employing the natural inner product

$$\langle f(x), g(x) \rangle = \int_0^1 f(x)g(x)\, dx, \tag{13.5}$$

we impose the Galerkin requirement

$$\langle x^i, R \rangle = 0 = \langle x^i, -1 \rangle + \sum_{j=1}^{N} c_j \langle x^i, j x^{j-1} + x^j \rangle \tag{13.6}$$

to obtain the N equations

$$\sum_{j=1}^{N} c_j \left(\frac{j}{i+j} + \frac{1}{i+j+1} \right) = \frac{1}{i+1}, \tag{13.7}$$

$i = 1, 2, 3, \ldots, N$. We solve these equations $Ac = b$ with a routine such as

```
% Routine 13.1  Galerkin problem
N = 5;                            % number of trial fns
                                  %
for i=1:N                         %
  b(i) = 1/(i+1);                 % entries of (row) b
  for j = 1:N                     %
   A(i,j) = j/(i+j)+ 1/(i+j+1); % entries of A
   end;                           %
  end;                            %
                                  %
c = A\b'                          % solve Ac = b' for column c
```

which solves for the coefficients c_j of the trial solution (13.3). We can graphically compare this approximate solution to the actual solution by adding several lines:

```
% Routine 13.1a  Graphing portion
x=0:0.05:1;              % x in steps of dx = 0.05
y = 1 - exp(-x);         % actual solution
d = -1;                  %
for j = 1:N              %
  d = - d*j;             %
  f(j) = 1/d;            % alternating reciprocal factorial
  X(:,j) = x'.^j;        % matrix of powers of x
end;
plot(x,y)                % plot actual solution
hold on                  % hold the actual soln.  plot
z = X*c;                 % the Galerkin approx soln.  at x
plot(x,z')               % plot approx.  soln.
w = X*f';                % N terms of Taylor series at x
plot(x,w','.')           % plot Taylor series approx.  as dots
```

The results of using $N = 2, \; 3, \; 4$ terms are shown in Figures 13.1a–13.1c. Notice the astonishing accuracy of the Galerkin method with even $N = 2$ terms. Even though both Galerkin and the Taylor series are employing powers of x, the Galerkin choice of coefficients is far more effective. The relative mean square error of Galerkin to Taylor

$$E(N) = \frac{\int_0^1 [\tilde{y}(x) - y(x)]^2 \, dx}{\int_0^1 [T_N(x) - y(x)]^2 \, dx} \tag{13.8}$$

(where $T_N(x)$ is the Nth-degree Taylor approximation), is approximately 1.5%, 0.14%, and 0.018% for $N = 2, 3, 4$ terms, respectively.

Sadly, this amazing Galerkin accuracy will not continue to improve as N increases without bound. With everything comes a price. The coefficient matrices of a Galerkin approximation are typically *ill-conditioned* for large N. The condition numbers of the coefficient matrix A of this problem are

$$\text{cond}(A) = \|A\|_2 \cdot \|A^{-1}\|_2 = 49, \; 1500, \; 47,000$$

for $N = 2, 3, 4$, respectively, requiring greater and greater care in solving for the coefficients c_j. Using Routine 13.1, accuracy will continue to improve over the Taylor series until $N \geq 8$, whereupon

MATLAB's careful equation solver can no longer hold off the numerical instabilities. For more on these issues, see [Golub and Ortega].

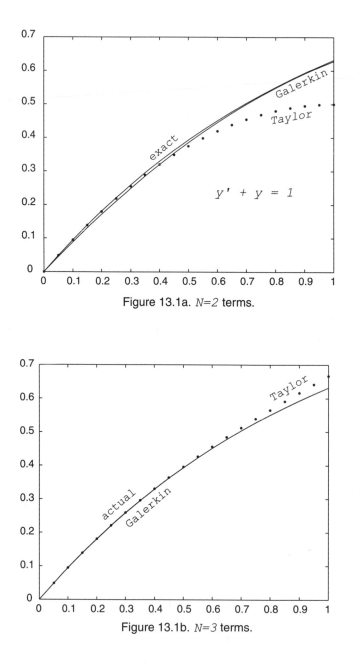

Figure 13.1a. $N=2$ terms.

Figure 13.1b. $N=3$ terms.

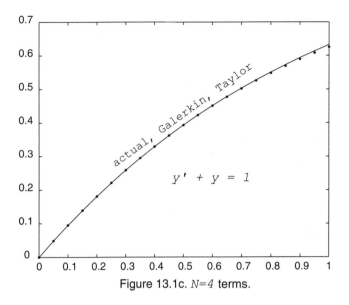

Figure 13.1c. $N=4$ terms.

To my knowledge the young Russian engineer B.G. Galerkin is the only person to do hard time for his mathematics — other Russian engineers became overconfident with Galerkin's method, using one- or two-term approximations in designing structures and systems that eventually failed. Stalin held Galerkin accountable.

§13.2 Eigenvalue Problems

The Galerkin approach transforms eigenvalue problems into the generalized eigenvalue problem $Ax = \lambda Bx$ of linear algebra.

Example 2. Let us solve the eigenvalue problem

$$y'' = \lambda y, \qquad 0 < x < 1, \qquad (13.9)$$

subject to zero boundary conditions $y(0) = 0 = y(1)$. Of course, we know the answer already from Problem 1 of Chapter 11: The eigenvalues are

$$\lambda_k = -k^2\pi^2 \text{ belonging to } y_k = \sin k\pi x. \qquad (13.10)$$

Let us employ the often-used basis functions

$$\phi_j = x^j - x^{j+1}, \qquad (13.11)$$

which obey the zero boundary conditions and (may) require no transcendental function support from the computer. The trial solution

$$\tilde{y} = \sum_{j=1}^{N} c_j \phi_j \tag{13.12}$$

of $Ly = y'' - \lambda y = 0$ has residual

$$R = \sum_{j=1}^{N} c_j \phi_j'' - \lambda \sum_{j=1}^{N} c_j \phi_j.$$

Imposing the Galerkin requirement yields the N equations

$$\sum_{j=1}^{N} c_j \langle \phi_i, \phi_j'' \rangle = \lambda \sum_{j=1}^{N} c_j \langle \phi_i, \phi_j \rangle, \tag{13.13}$$

i.e., the *generalized eigenvalue problem*

$$Sc = -\lambda Gc, \tag{13.14}$$

where G is the *Gramian*

$$G = (\langle \phi_i, \phi_j \rangle) \tag{13.15}$$

and S the *Sobolev* matrix

$$S = (\langle \phi_i', \phi_j' \rangle) = -(\langle \phi_i, \phi_j'' \rangle). \tag{13.16}$$

The last equality follows by integration by parts since the ϕ_k satisfy zero boundary conditions. By Exercise 13.3,

$$G = \left(\frac{1}{i+j+1} - \frac{2}{i+j+2} + \frac{1}{i+j+3} \right) \tag{13.17}$$

and

$$S = \left(\frac{ij}{i+j-1} - \frac{2ij+i+j}{i+j} + \frac{(i+1)(j+1)}{i+j+1} \right). \tag{13.18}$$

The generalized eigenvalue problem (13.14) yields easily to Routine 13.2:

```
% Routine 13.2  Galerkin eigenvalue problem
N= 5;                      %
for i = 1:N                %
 for j = 1:N               % define G and S:
  G(i,j) = 1/(i+j+1)-2/(i+j+2) + 1/(i+j+3);
  S(i,j) = i*j/(i+j-1)-(2*i*j+i+j)/(i+j)+(i+1)*(j+1)/(i+j+1);
 end;                      %
 end;                      %
                           %
[X,D] = eig(S,-G);         %
D                          %
```

yielding the estimates for the first five eigenvalues $\lambda_k = -k^2\pi^2$ shown in Table 13.1.

Table 13.1

k	1	2	3	4	5
Galerkin	-9.8696	-39.5016	-89.1736	-200.9568	-350.9568
Actual	-9.8696	-39.4784	-88.8264	-157.9136	-246.7401

Plotting the first few approximate eigenfunctions is left as Exercise 13.5.

By a fundamental insight of H. Weyl, called the *maximum minimum principle*, the approximate eigenvalues will always be overestimates [Courant and Hilbert, vol. 1, p. 408].

§13.3 Steady Problems

Example 3. Let us solve the Poisson problem

$$\nabla^2 u = -1, \qquad 0 < x,\ y < 1, \qquad (13.19)$$

on the unit square subject to zero boundary conditions — a model of steady fluid flow down a square tube (Exercise 11.34). We attempt the trial solution

$$\tilde{u} = \sum_{i,j=1}^{N} c_{ij}\phi_{ij}(x,y) \qquad (13.20)$$

using the basis functions

$$\phi_{ij}(x,y) = \alpha_i(x)\beta_j(y) = (x^i - x^{i+1})(y^j - y^{j+1}). \qquad (13.21)$$

The residual for this trial solution is

$$R = 1 + \nabla^2 \tilde{u} = 1 + \sum_{i,j=1}^{N} c_{ij} \nabla^2 \phi_{ij}. \qquad (13.22)$$

Imposing Galerkin's requirement,

$$0 = \langle \phi_{pq}, 1 + \nabla^2 \tilde{u} \rangle = \langle \phi_{pq}, 1 \rangle + \sum_{i,j=1}^{N} c_{ij} \langle \phi_{pq}, \nabla^2 \phi_{ij} \rangle, \qquad (13.23)$$

i.e., the N^2 equations

$$\sum_{i,j=1}^{N} c_{ij} \langle \alpha_p \beta_q, \alpha_i'' \beta_j + \alpha_i \beta_j'' \rangle = -\langle \phi_{pq}, 1 \rangle, \qquad (13.24)$$

$p, q = 1, 2, \ldots, N$ in the N^2 unknowns c_{ij}, i.e.,

$$Sc = (\langle \phi, 1 \rangle).$$

In Exercise 13.8 you are asked to solve this system and compare the Galerkin approximate solution to the exact solution obtained analytically. Especially compare the accuracy for the single-term approximation $\tilde{u} = 1.25(x - x^2)(y - y^2)$. You will be amazed.

§13.4 Transient Problems

For transient problems, say a diffusion problem

$$u_t = \nabla^2 u,$$

the trial solution is taken to have time-varying coefficients

$$\tilde{u} = \sum_{j=1}^{N} c_j(t) \phi_j. \qquad (13.25)$$

Galerkin's requirement on the residual $R = \tilde{u}_t - \nabla^2 \tilde{u}$ in this case yields

$$\sum_{j=1}^{N} \dot{c}_j(t) \langle \phi_i, \phi_j \rangle = \sum_{j=1}^{N} c_j(t) \langle \phi_i, \nabla^2 \phi_j \rangle, \qquad (13.26)$$

a system of N ODEs in the N variables $c_j = c_j(t)$. The initial conditions for this system are obtained by imposing the Galerkin requirement on the residual of the initial condition of the PDE, i.e., insist that

$$0 = \langle \phi_i, u(x,0) - \tilde{u}(x,0) \rangle,$$

i.e.,

$$\sum_{j=1}^{N} c_j(0)\langle \phi_i, \phi_j \rangle = \langle \phi_i, u(x,0) \rangle. \tag{13.27}$$

Solve the system (13.26) with a numerical ODE solver subject to the initial conditions obtained by solving (13.27).

Example 4. Let us numerically solve the wave equation

$$u_{tt} = u_{xx}, \qquad 0 < x < 1, \tag{13.28}$$

subject to zero boundary conditions $u(0,t) = 0 = u(1,t)$ and subject to the initial condition $u(x,0) = 0$ and terminal condition $u(x,1) = 1$.
  Our trial solution will be

$$\tilde{u}(x,t) = \sum_{j=1}^{N} c_j(t)\phi_j(x), \tag{13.29}$$

where we employ as usual the basis functions

$$\phi_j(x) = x^j - x^{j+1}. \tag{13.30}$$

Imposing the Galerkin requirement on the residual of the PDE applied to $\tilde{u}$ yields the second-order system

$$\sum_{j=1}^{N} \ddot{c}_j(t)\langle \phi_i, \phi_j \rangle = \sum_{j=1}^{N} c_j(t)\langle \phi_i, \phi_j'' \rangle, \tag{13.31}$$

i.e.,

$$G\ddot{c} = -Sc. \tag{13.32}$$

  This vibrational problem (13.32) yields to the generalized eigenvalue problem. Because the Gram and Sobolev matrices G and S are symmetric and positive definite (Exercise 13.7), by the spectral theorem [Brown] the eigenvalues $\lambda_k = -\omega_k^2$ are negative and the corresponding eigenvectors v_k span. Thus the general solution to (13.32) is

$$c(t) = \sum_{k=1}^{N}(d_k e^{i\omega_k t} + e_k e^{-i\omega_k t})v_k$$

$$= \sum_{k=1}^{N}(a_k \cos \omega_k t + b_k \sin \omega_k t)v_k. \tag{13.33}$$

The eigenvectors v_k and their values λ_k are obtainable via MATLAB with the single request [V,D] = eig(S,-G) as in Routine 13.2. It remains to impose the initial and terminal conditions

$$\langle \phi_i, \tilde{u}(x,0) - 0 \rangle = 0 \tag{13.34a}$$

and

$$\langle \phi_i, \tilde{u}(x,1) - 1 \rangle = 0; \tag{13.34b},$$

i.e., solve

$$Gc(0) = 0 \tag{13.35a}$$

and

$$Gc(1) = (\langle \phi_i, 1 \rangle). \tag{13.35b}$$

Once $c(0)$ and $c(1)$ are determined, solve (13.33) for a_k and b_k (Exercise 13.9).

§13.5 Finite Elements

Some spatial domains are awkwardly shaped. Moreover a physical process may have more "activity" in certain portions of the domain, say near a boundary point where a steep change in temperature occurs, or near a corner where shear stresses are high. The approach is then to choose basis functions ϕ_j with small *support*, i.e., functions that are nonzero only on small subdomains of the spatial domain of the problem. Concentrate many of these basis functions where the action is taking place, few where activity is low. By these means, accuracy is improved, *and* the matrices of the numerical model become sparse.

Example 5. Let us numerically solve

$$y'' + y = -1, \qquad 0 < x < 1, \tag{13.36}$$

subject to the zero boundary conditions $y(0) = 0 = y(1)$. The actual solution is (Exercise 13.10)

$$y(x) = -1 + \cos x + \frac{1 - \cos 1}{\sin 1} \sin x. \tag{13.37}$$

Divide the interval $[0,1]$ into N equal subintervals, each of length $\Delta x = 1/N$.

For $j = 1, 2, 3, \ldots N - 1$, take the piecewise linear basis function ϕ_j that is zero off the open interval $((j-1)\Delta x, (j+1)\Delta x$ but has value 1 at $j\,\Delta x$ as shown in Figure 13.2.

Figure 13.2. Basis functions with small support.

Attempt the piecewise linear trial solution

$$\tilde{y} = \sum_{j=1}^{N-1} c_j \phi_j \tag{13.38}$$

and impose the Galerkin requirement on the residual of $Ly = y'' + y + 1$ to obtain

$$\sum_{j=1}^{N-1} c_j \langle \phi_i, \phi_j'' + \phi_j \rangle$$

$$= \sum_{j=1}^{N-1} c_j \langle \phi_i, \phi_j'' \rangle + c_j \langle \phi_i, \phi_j \rangle = -\langle \phi_i, 1 \rangle. \tag{13.39}$$

But something has gone wrong. This piecewise linear interpolation (13.38) has zero second derivative almost everywhere, so our purported equations (13.39) reduce to simply

$$Gc = -(\langle \phi_i, 1 \rangle),$$

which instead finds the best fit to the horizontal line $y = 1$; it is not solving the differential equation (13.36). This same failure will always occur when using piecewise linear trial functions ϕ_j to solve second-order problems.

One way to view this failure is that the approach does not take into account the Dirac delta functions that should arise when differentiating these basis functions ϕ_j twice. Another (equivalent) view is that we must not ask so much of solutions — they need not be so differentiable. Rather than requiring the solution satisfies the classical statement of the problem

$$y''(x) + y(x) = -1, \tag{13.40}$$

we only require that the solution holds when projected into finite-dimensional subspaces, i.e., satisfies the weak condition

$$-\langle \phi', y' \rangle + \langle \phi, y \rangle = -\langle \phi, 1 \rangle \tag{13.41}$$

for any *test* function ϕ with square-integrable derivative and zero boundary values. In short, by integrating by parts, we *throw one derivative onto the test function*. Thus the weak restatement of our Galerkin problem (13.39) is

$$-\sum_{j=1}^{N-1} c_j \langle \phi_i', \phi_j' \rangle + \sum_{j=1}^{N-1} c_j \langle \phi_i, \phi_j \rangle = -\langle \phi_i, 1 \rangle, \tag{13.42}$$

$i = 1, 2, 3, \ldots, N - 1$, i.e.,

$$(G - S)c = -(\langle \phi_i, 1 \rangle). \tag{13.43}$$

Because all the basis functions ϕ_j are translates of one another, G and S are easy to hand compute (Exercise 13.12), a hallmark of such finite element problems:

$$\langle \phi_i, \phi_j \rangle = \begin{cases} 2/3N & \text{if } i = j \\ 1/6N & \text{if } |i - j| = 1 \\ 0 & \text{otherwise,} \end{cases} \tag{13.44}$$

$$\langle \phi_i', \phi_j' \rangle = \begin{cases} 2N & \text{if } i = j \\ -N & \text{if } |i - j| = 1 \\ 0 & \text{otherwise,} \end{cases} \tag{13.45}$$

and

$$\langle \phi_i, 1 \rangle = \frac{1}{N}. \tag{13.46}$$

Thus (13.43) becomes the tridiagonal system

$$
\frac{1}{6N}
\begin{pmatrix}
4 & 1 & 0 & . & . & 0 \\
1 & 4 & 1 & 0 & . & 0 \\
. & 0 & . & . & & . \\
. & . & & & . & 1 \\
0 & . & . & 0 & 1 & 4
\end{pmatrix}
\begin{vmatrix}
c_1 \\
c_2 \\
. \\
. \\
c_{N-1}
\end{vmatrix}
$$

$$
-N
\begin{pmatrix}
2 & -1 & 0 & . & . & 0 \\
-1 & 2 & -1 & 0 & . & 0 \\
. & 0 & . & . & & . \\
. & . & & & . & -1 \\
0 & . & . & 0 & -1 & 2
\end{pmatrix}
\begin{vmatrix}
c_1 \\
c_2 \\
. \\
. \\
c_{N-1}
\end{vmatrix}
= -\frac{1}{N}
\begin{vmatrix}
1 \\
1 \\
. \\
. \\
1
\end{vmatrix},
$$

implemented as follows:

```
% Routine 13.3  y'' + y = -1
N= 6;                        %
Id = eye(N-1);               % identity matrix
T = ones(N-1);               % matrix of all 1's
T = tril(T,1);               % lower triangular
T = triu(T,-1);              % now tridiagonal(1,1,1)
T = T - Id;                  % now tridiagonal(1,0,1)
G = 2*Id/(3*N)+T/(6*N);      % the Gramian
S = 2*N*Id - N*T;            % the Sobolev matrix
b = - ones(N-1,1)/N;         % column of constants
                             %
c = (G-S)\b;                 % solves the system
```

To view the resulting accuracy of this Galerkin linear interpolation, first add the canonical basis element shown in Figure 13.3 to MATLAB's library by saving the script

```
function y = phi(x)
y = (1-sign(x*x-1))*(1-abs(x))/2
```

as the file **phi.m**.

Our basis elements ϕ_j used in our Galerkin trial solution (13.38) are translations and dilations of this canonical ϕ, shown in Figure 13.3, namely $\phi_j(x) = \phi(Nx - j)$, and can be used as such in a graphing script. A graphing routine will yield Figures 13.4a and 13.4b.

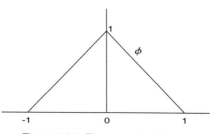

Figure 13.3. The canonical basis
function from which all others are obtained.

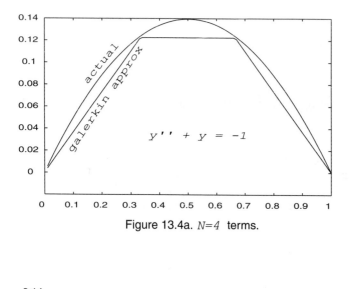

Figure 13.4a. *N=4* terms.

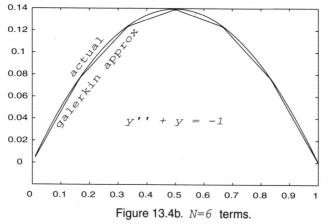

Figure 13.4b. *N=6* terms.

Problem. Find the vibrational frequencies ω_k of a triangular drum.

Solution. We must numerically investigate the wave equation

$$u_{tt} = \nabla^2 u, \qquad \Omega : 0 < x + y < 1, \ 0 < x, y, \qquad (13.47)$$

subject to zero boundary conditions. This is a good test project since we will be able to check our numerical results against exact analytic results — each vibrational mode of this $45°$ triangular drum can be extended to a mode of the square drum (Exercise 11.21).

By Rellich (Remark 1 of §11.3), all solutions to (13.47) are of the form

$$u(x, t) = \sum_{k=1}^{\infty} a_k \cos(\omega_k t - \theta_k)\psi_k(x, y)$$

where ψ_k is the eigenfunction belonging to the kth eigenvalue $\lambda_k = -\omega_k^2$ of the Laplacian

$$\nabla^2 \psi_k = \lambda_k \psi_k \qquad (13.48)$$

subject to zero boundary conditions on this triangular domain Ω. These $\omega_k = \sqrt{-\lambda_k}$ are the natural vibrational frequencies that combine in some combination of energy distribution to form the sounds of such a drum.

We go by finite elements, with continuous piecewise linear basis functions. We disjoint our triangular domain Ω into many smaller triangular domains Ω_k, as in Figure 13.5.

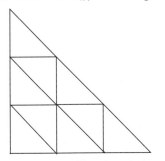

Figure 13.5. The domain is
subdivided into N subdomains.

Each basis element ϕ_j will be 1 at some vertex but zero at all other vertices, forming a "tepee"-shaped surface as in Figure 13.6.

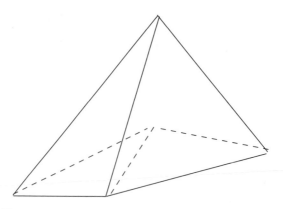

Figure 13.6. Each basis element is a tepee-shaped
surface of height 1 above one vertex, 0 at all others.

We exchange our classical problem $\nabla^2 u = \lambda u$ for the weak re-statement

$$-[\nabla\phi, \nabla u] = \lambda\langle\phi, u\rangle \qquad (13.49)$$

and impose the Galerkin requirement on the trial solution

$$\tilde{u} = \sum_{j=1}^{N} c_j \phi_j \qquad (13.50)$$

to obtain the numerical model, the generalized eigenvalue problem

$$Sc = -\lambda Gc, \qquad (13.51)$$

where as usual

$$G = (\langle\phi_i, \phi_j\rangle), \qquad (13.52)$$

$$S = ([\phi_i, \phi_j]), \qquad (13.53)$$

and where the *Dirichlet inner product*

$$[\phi, \psi] = \int_\Omega \nabla\phi \circ \nabla\psi \, d\Omega. \qquad (13.54)$$

The Dirichlet inner product $[\cdot, \cdot]$ is merely the multidimensional version of $\langle\phi', \psi'\rangle$. I leave the remaining work as a major student project (Exercise 13.15).

§13.6* Why So Effective?

You must have asked by now why Galerkin's method is so effective, why such extraodinary accuracy is achieved with modest effort. It is a tale of two norms. If u is the exact solution to a problem $Lu = f$, $\tilde{u}$ a Galerkin approximate solution, and $e = u - \tilde{u}$ the error, then as a general rule

$$\langle e, e \rangle \leq \eta^2 \cdot (e, e), \tag{13.55}$$

where $(\phi, \psi) = \langle L\phi, \psi \rangle$ is a physically natural (potential) energy inner product associated with the problem. The Galerkin requirement yields an approximate solution $\tilde{u}$ with error e that minimizes the potential energy (e, e) (Exercise 13.19), thus forcing the mean square error $\langle e, e \rangle$ small.

The rate of convergence $e \to 0$ with N is estimated from above by using some convenient (rather than optimal) choice of coefficients of the trial solution

$$\tilde{y} = \sum_{j=1}^{N} c_j \phi_j.$$

Example 6. Let us consider the Poisson problem $\nabla^2 u = -f$ on a bounded planar domain Ω subject to zero boundary conditions. Physically, we are solving for the vertical displacement $u = u(x, y)$ of a drumhead stretched over Ω that is subjected to the distributed vertical load f. When vibrating, the displacement is governed by

$$u_{tt} = \nabla^2 u,$$

the left-hand side representing the inertial force $F = ma$, the right-hand side representing the restoring force of the elastic drumhead. Thus summing up distributed force $-\nabla^2 u$ times displacement u,

$$-\int_\Omega u \nabla^2 u \, d\Omega = -\langle u, \nabla^2 u \rangle = \int_\Omega \nabla u \circ \nabla u \, d\Omega = [u, u] \tag{13.56}$$

represents (within units) the total work needed to distort the drumhead into the configuration u.

The natural (*Hilbert*) inner product for this problem is

$$\langle \phi, \psi \rangle = \int_\Omega \phi \cdot \psi \, d\Omega \tag{13.57}$$

while the potential energy (Dirichlet) inner product is

$$[\phi, \psi] = \int_\Omega \nabla\phi \circ \nabla\psi \, d\Omega \qquad (13.58)$$

(see Exercise 13.21).

The energy norm *dominates* the usual norm, i.e.,

$$\langle\phi, \phi\rangle \le \eta^2 [\phi, \phi], \qquad (13.59)$$

(where η is the minimal width of Ω) for any function ϕ with zero boundary values and square-integrable first derivatives on Ω (Exercise 13.18), giving the inequality (13.55). In particular, if $\phi = \phi_1$ is the eigenfunction belonging to the dominant eigenvalue λ_1 of the Laplacian, then

$$\frac{[\phi_1, \phi_1]}{\langle\phi_1, \phi_1\rangle} = -\frac{\langle\nabla^2\phi_1, \phi_1\rangle}{\langle\phi_1, \phi_1\rangle} = -\lambda_1 \ge \frac{1}{\eta^2}, \qquad (13.60)$$

so *the lowest (nondimensional) frequency* $\omega = \sqrt{-\lambda}$ *sounded by a drum is at the very least the reciprocal of the shortest width of the drum.* The smaller the drum, the higher the sound.

Let us now triangulate the domain Ω and employ the piecewise linear tepee-shaped basis functions ϕ_j that are 1 above one vertex, zero at all others, as in Figure 13.6. We then consider the geodesic dome–like finite element approximations

$$\tilde{u} = \sum_{j=1}^{N} c_j\phi_j \qquad (13.61)$$

to the shape of our drumhead u. We must at this point assume that the boundary is piecewise smooth and convex so that none of the triagular sides lie outside the closure $\overline{\Omega}$ of the domain Ω. Thus each approximation $\tilde{u}$ vanishes off Ω, allowing integration by parts (Exercise 11.6).

Impose the Galerkin requirement to select an approximation $\tilde{u}$ to u of form (13.61) with error $e = u - \tilde{u}$. Thus, as in Exercise 13.20,

$$[e, e] = \langle f, e\rangle. \qquad (13.62)$$

Our goal is to show that

$$\|e\| = \sqrt{\langle e, e\rangle} \le \rho\|f\|h^2, \qquad (13.63)$$

where h is the longest side length of the triangles of our triangulation. To accomplish this goal, we employ the terminally clever *Nitsche's trick* [Strang and Fix] that bounds the Galerkin derived error e by interpolation error:

Step 1. Solve the Poisson problem $\nabla^2 v = -e$ subject to zero boundary conditions on Ω.

Step 2. Linearly interpolate the solution v with $\tilde{v}$ of form (13.61) so that v and $\tilde{v}$ agree at each vertex of the triangulation.

Step 3. Note that $[v, w] = \langle e, w \rangle$ for all smooth w that vanish on the boundary $\partial\Omega$, and in particular because of the Galerkin orthogonality requirement,

$$\langle e, e \rangle = [v, e] = [v - \tilde{v}, e]. \tag{13.64}$$

Hence by Cauchy's inequality,

$$\langle e, e \rangle^2 = [v - \tilde{v}, e]^2 \leq [v - \tilde{v}, v - \tilde{v}] \cdot [e, e]$$

$$= \langle e, v - \tilde{v} \rangle \cdot \langle f, e \rangle \leq \|e\|^2 \|f\| \cdot \|v - \tilde{v}\|,$$

giving

$$\langle e, e \rangle \leq |f| \cdot \|v - \tilde{v}\|. \tag{13.65}$$

Step 4. Apply the standard estimate for the accuracy of linear interpolation [Prenter]:

$$\|v - \tilde{v}\| \leq Mh^2, \tag{13.66}$$

where M is determined by the second partial derivatives of v:

$$M = \int_\Omega \sum_{i,j} (D_{i,j}v)^2 \, d\Omega.$$

Step 5. Show that M is bounded by its data e [Strang and Fix]:

$$M \leq \rho \|e\|, \tag{13.67}$$

and thus by combining (13.65)–(13.67), the Galerkin error is at worst

$$\|e\| \leq \rho \|f\| h^2. \tag{13.68}$$

Exercises

13.1 Compare the accuracy and effort required of a divided difference solution of the problem of Example 1 versus the Galerkin
solution.

13.2 Solve the problem of Example 1 using the basis functions $\phi_0 = x$, $\phi_j = x^j - x^{j+1}$ for $j = 1, 2, \ldots, N$.

13.3 Establish the formulas for G and S in (13.17) and (13.18).

13.4 Numerically solve for the solution of *Airy's equation* $y'' = xy$
on the interval $0 < x < 1$ with boundary conditions $y(0) = y'(0) = 1$. Use the trial solution

$$\tilde{y} = 1 + x + \sum_{j=2}^{N} c_j x^j.$$

13.5 Plot the first several Galerkin approximate eigenfunctions obtained from Routine 13.2 against the actual $y_k = \sin k\pi x$.

13.6 Numerically estimate the first six vibrational frequencies of a
round drum that is struck in the center. Use the basis $\phi_j = r^j - r^{j+1}$ and the natural inner product for this problem (11.54).
Compare with the actual (nondimensional) values α_k where
$J_0(\alpha_k) = 0$ (see Table on page 206).

13.7* Show that for any N elements $\psi_1, \psi_2, \ldots, \psi_N$ of an inner product space, the Gram matrix $A = (\langle \psi_i, \psi_j \rangle)$ is positive semidefinite. If the ψ_i are linearly independent, then A is positive
definite (and conversely).

13.8 Solve (13.19) analytically by using the homogeneous spatial
modes

$$u(x, y) = \sum_{m,n=1}^{\infty} c_{ij} \sin m\pi x \; \sin n\pi y$$

and solving for the c_{ij}. Modify Routine 13.1 to handle the system (13.24). Compare the accuracy of the Galerkin approximate solution to the exact solution.

13.9 (**Project**) Supply all missing details in the Galerkin approximate solution to (13.28), including a graphical comparison with
the actual solution.

13.10 Find a Galerkin approximate solution to the diffusion problem (11.16) using the basis functions $\phi_j = x^j - x^{j+1}$.

13.11 Solve (13.36) analytically.

13.12 Verify the entries (13.44) of G and the entries (13.45) of S.

13.13 Write a Fortran or C subroutine to solve any invertible tridiagonal system of equations using at most 16 lines of code.

13.14 Find the first few axial vibrational frequencies of a rod fixed at each end whose density increases linearly; i.e., solve $xu_{tt} = u_{xx}$ subject to $u(0,t) = 0 = u(1,t)$.

13.15 (**Project**) Complete the work needed to solve (13.48) for the first few vibrational frequencies of the triangular drum.

13.16 Model the diminishing axial vibrations of a fixed homogeneous rod with viscous damping, i.e., $u_{tt} + 0.01u_{tx} - u_{xx} = 0$ with zero boundary conditions $u(0,t) = 0 = u(1,t)$ and initial condition $u(x,0) = \sin \pi x$, $u_t(x,0) = 0$.

13.17 (**Project**) Estimate via Galerkin the dominant normal vibrational frequency of a thin circular disk fixed at its center. Model normal displacement with the plate equation $u_{tt} + \nabla^2\nabla^2 u = 0$. Assume displacement at this dominant mode is radially symmetric, i.e., $u = u(r,t)$. Formulate the four physically natural boundary conditions. Obtain an analytic solution by separation of variables.

13.18* Prove the *Poincaré*-like inequality $\langle \phi, \phi \rangle < c^2 [\phi, \phi]$ for all functions ϕ with square-integrable derivatives over the domain Ω and zero boundary values, where c is the least width of Ω.

Outline: We may assume that the first coordinate x_1 of points in Ω lies between the limits $a_1 < x_1 < b_1$ with $c = b_1 - a_1$. Extend ϕ to be zero off Ω. Then by the fundamental theorem of calculus and Cauchy's inequality,

$$\phi(x_1, x_2, \dots, x_n)^2 = [\int_{a_1}^{x_1} \phi_{x_1}(t, x_2, \dots, x_n)\, dt]^2$$

$$\leq c \int_{a_1}^{x_1} \phi_{x_1}(t, x_2, \dots, x_n)^2\, dt.$$

Thus

$$\langle \phi, \phi \rangle = \int_\Omega \phi^2 \, d\Omega \;\leq\; c^2 \int_\Omega \phi_{x_1}^2 \, d\Omega \;\leq\; c^2 [\phi, \phi].$$

13.19*Let u be an element of an inner product space X with (positive definite) inner product $(\cdot, \cdot)$. Suppose S is a finite subspace. Then if $\tilde{u}$ is chosen from S so that $e = u - \tilde{u}$ is orthogonal to S, then

$$(e, e) = \min_{v \in S}(u - v, u - v).$$

Hint: Expand out $(e + \epsilon v, e + \epsilon v)$.

13.20*Under the notation in force from Exercise 13.19, prove that if $Lu = f$, then $\langle Lu, e \rangle = (e, e) = \langle f, e \rangle$, so the inequality $\langle e, e \rangle \leq \eta(e, e)$ implies $(e, e) \leq \eta^2 \langle f, f \rangle$.

13.21 Show that in any vibrational problem $\ddot{x} = Ax$, where A is self-adjoint, that total energy $H = \langle \dot{x}, \dot{x} \rangle - \langle Ax, x \rangle$ is constant along trajectories. Show that in common examples the term $V = -\langle Ax, x \rangle / 2$ is (within a constant multiple) potential energy.

13.22*(Project) Obtain a tight estimate of the Galerkin error $\langle e, e \rangle$ for Example 1 of §13.1 (assuming infinite percision) that is consistent with the data obtained from Routine 13.1.

13.23 (Project) Attempt to best MATLAB's equation solver. Using large (ill-conditioned) Hilbert matrices $A = (1/(i + j))$, select a vector x_0 and set $b = Ax_0$. Write a routine in C or Fortran to solve the system $Ax = b$ for x and compare with MATLAB's answer x = A\b. Compare your accuracy against the truth model x_0 with various norms, e.g., largest error in any one coordinate ($\|x - x_0\|_\infty$), the sum of absolute errors over all components ($\|x - x_0\|_1$), or especially, the square root of the sum of the squares of component errors ($\|x - x_0\|_2$). Can you achieve even 20% of MATLAB's accuracy?

Chapter 14

Splines

The mathematical analog of the flexible drafting tool is the cubic spline.

Piecewise linear approximations cannot be used for fourth-order finite element solutions, and as we see, nor can quadratics. We must turn to cubics. Special piecewise polynomial functions called *splines* are introduced to smoothly interpolate discrete data. Cubic splines are shown to solve the natural interpolation problem. A special basis of cubic B-splines is revealed and used to find the shape of an earth retaining wall.

§14.1 Why *Cubics?*

We return to the problem of approximating a function u on a domain Ω by linear combinations of basis functions

$$\psi = c_1\phi_1(x) + c_2\phi_2(x) + \cdots + c_N\phi_N(x). \tag{14.1}$$

In particular, we are interested in approximating solutions to fourth order problems such as

$$\nabla^2\nabla^2 u = g \tag{14.2}$$

subject to zero, and zero flux boundary conditions. This models the deflection u of a beam or plate clamped at its boundary, loaded by distributed forces g.

Imposing the Galerkin requirement (13.2) on ψ of (14.1) as a trial solution of (14.2) and integrating twice by parts yields the weak reformulation

$$\sum_{j=1}^{N} c_j\langle\nabla^2\phi_j, \nabla^2\phi_i\rangle = \langle g, \phi_i\rangle, \tag{14.3}$$

$i = 1, 2, \ldots, N$, thus yielding N equations in the N unknowns c_j.

In contrast to the first- and second-order problems solved in Chapter 13 with finite elements, we cannot use piecewise linear approximations ψ in the weak reformulation (14.3), since the left side becomes 0 and Galerkin's requirement cannot be satisfied. One might then guess that piecewise quadratic approximations will work. They will not! A heuristic argument to this effect follows:

To keep the ideas simple, let us cut back to one spatial dimension. Suppose that we are interested in the lateral deflection $y = y(x)$ of an earth retaining wall at increasing depth x, modeled by $y'''' = kxy$ subject to $y(a) = y'(a) = 0 = y(b) = y'(b)$.

We employ a *piecewise polynomial* trial function $y = \psi(x)$ on $[a, b]$; this means that the interval $[a, b]$ can be partitioned into n subintervals $a = x_0 < x_1 < \cdots < x_n = b$, whereupon ψ is given on each subinterval (x_{i-1}, x_i) by a polynomial ψ_i in x.

When we attempt to throw derivatives from a twice-differentiable function η onto ψ by integrating by parts,

$$\langle \eta'', \psi \rangle = \eta'(x)\psi(x)|_a^b - \langle \eta', \psi' \rangle, \tag{14.4}$$

the *boundary term*

$$\eta'(x)\psi(x)|_a^b = \sum_{i=1}^{n} \eta'(x_i)\psi_i(x_i^+) - \eta'(x_{i-1})\psi_i(x_{i-1}^-) \tag{14.5}$$

cannot telescope to

$$\eta'(b)\psi(b) - \eta'(a)\psi(a) = 0 \tag{14.6}$$

unless the polynomial elements fit together continuously; i.e., $\psi_i(x_i^+) = \psi_{i+1}(x_i^-)$. Assuming such continuity at each *knot* x_i, upon integrating again by parts,

$$\langle \eta', \psi' \rangle = \eta(x)\psi'(x)|_a^b - \langle \eta, \psi'' \rangle, \tag{14.7}$$

we see that the boundary term again will not vanish unless the *derivatives* of the polynomial elements fit together continuously.

For a piecewise polynomial function ψ to be an admissible approximation in a fourth-order problem, it must be continuously differentiable.

Thus for piecewise quadratics to qualify as approximations in fourth-order problems they must be 2-*splines*, i.e., continuously differentiable.

But the 2-splines cannot be dense and at the same time satisfy four end conditions — there are simply too many conditions to satisfy: the $3n$ coefficients of the n quadratic elements ψ_i of ψ must satisfy typically $2n$ interpolation requirements $y_{i-1} = \psi_i(x_{i-1})$, $y_i = \psi_i(x_i)$, plus $n-1$ interior smoothness conditions $\psi_i'(x_i) = \psi_{i+1}'(x_i)$, plus two additional end conditions. This amounts to $3n+1$ conditions imposed on $3n$ degrees of freedom. Quadratics will not work. We must employ piecewise cubic trial functions.

§14.2 *m*-Splines

DEFINITION. An *m-spline* is a function with $m-1$ continuous derivatives that is given piecewise by polynomials of degree at most m.

The symbol
$$S^m(\Pi) \tag{14.8}$$

will denote the real vector space of all m-splines on the interval $[a,b]$ partitioned by $\Pi : a = x_0 < x_1 < x_2 < \cdots < x_n = b$. The symbol $S_0^m(\Pi)$ will denote the subspace of $S^m(\Pi)$ of all ψ with zero end values $\psi(a) = 0 = \psi(b)$.

Note that if we had asked an m-spline ψ to have m rather than $m-1$ continuous derivatives, the knots would disappear — ψ would be a single polynomial of degree at most m throughout the interval $[a,b]$ (Exercise 14.2).

RESULT A. Let n be the number of subintervals of the partition Π of $[a,b]$. Then
$$\dim S^m(\Pi) = n + m \tag{14.9}$$

and
$$\dim S_0^m(\Pi) = n + m - 2. \tag{14.10}$$

PROOF. Note that differentiation yields a surjective map
$$D : S^m(\Pi) \to S^{m-1}(\Pi) \tag{14.11}$$

since
$$\int_0^x \; : S^{m-1}(\Pi) \to S^m(\Pi).$$

The linear operator D has as kernel the constant functions, so differentiation lowers rank by 1 in (14.11). But $S^0(\Pi)$, the space of all piecewise constant functions on the partition Π, clearly has rank n, thus establishing (14.9) by induction.

The subspace $S_0^m(\Pi)$ of $S^m(\Pi)$ is the intersection of two hyperspaces, the kernels of the point evaluation linear functionals $\zeta(\psi) = \psi(a)$ and $\eta(\psi) = \psi(b)$. These two linear functionals have distinct kernels; for instance, they disagree at a polynomial of degree m or less that vanishes at a but not at b. Thus $S_0^m(\Pi)$ is of rank 2 less than $S^m(\Pi)$.

Example 1. The 1-splines are piecewise linear and continuous. They can interpolate any function f at the $n+1$ knots $a = x_0 < x_1 < \cdots < x_n = b$ by simply connecting the $n + 1$ points $(x_i, f(x_i))$ by straight-line segments. A basis of $S^1(\Pi)$ of elements of smallest support consists of two-piece functions ϕ_i, shown in Figure 14.1. Strike out ϕ_0 and ϕ_n to obtain a basis of $S_0^1(\Pi)$.

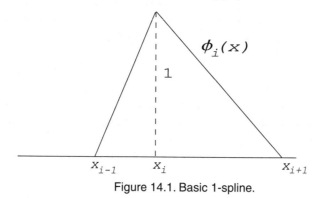

Figure 14.1. Basic 1-spline.

Example 2. The quadratic splines can interpolate any function f at the $n+1$ knots $a = x_0 < \cdots < x_n = b$ in a continuously differentiable manner. However, it is generically not possible to interpolate the derivatives $f'(x_i)$ (Exercise 14.3). A basis of elements with minimum support for the case of equal subdivisions $h = x_{i+1} - x_i$ is given by the $n + 2$ translates $i = -1, \ldots, n$ of the three-piece function ϕ_i shown in Figure 14.2. An analytic formula for this function is left as Exercise 14.4.

It is understood that these translates are zero off $[a, b]$. For a basis of $S_0^m(\Pi)$, discard $\phi_{-1}, \phi_0, \phi_{n-1}, \phi_n$ and replace with two carefully chosen 2-splines (Exercise 14.5).

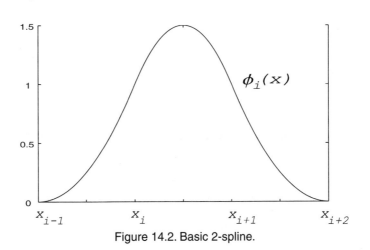

Figure 14.2. Basic 2-spline.

Experts in computer-assisted drawing (CAD) tell me that the human eye is offended by the lack of continuous second derivatives when interpolating with 2-splines. Even 3-splines fall short when rendering reflections.

§14.3 Cubic Splines

Most useful by far are the cubic splines. They are capable of interpolating the $n + 1$ values $f(x_i)$, $0 \leq i \leq n$, plus meeting two additional conditions[1] — see Result B below and Exercise 14.6. For equally spaced knots, a basis of $S^3(\Pi)$ consisting of elements of minimum support is given by the $n + 3$ translates, $i = -1, \dots, n + 1$, of the four-piece *cubic B-spline* B_i, shown in Figure 14.3.

Explicitly, if $h = x_i - x_{i-1} = (b - a)/n$, then take

$$B_i(x) = \frac{1}{h^3} \begin{cases} (x - x_{i-2})^3 & \text{if } x_{i-2} \leq x \leq x_{i-1} \\ (h + x - x_{i-1})^3 - 4(x - x_{i-1})^3 & \text{if } x_{i-1} \leq x \leq x_i \\ (h + x_{i+1} - x)^3 - 4(x_{i+1} - x)^3 & \text{if } x_i \leq x \leq x_{i+1} \\ (x_{i+2} - x)^3 & \text{if } x_{i+1} \leq x \leq x_{i+2} \\ 0 & \text{otherwise} \end{cases}$$

$$(14.12)$$

---

[1] The $n+1$ interpolations plus the $2(n-1)$ first and second continuous derivative requirements at interior knots use up all but 2 of the $4n$ degrees of freedom.

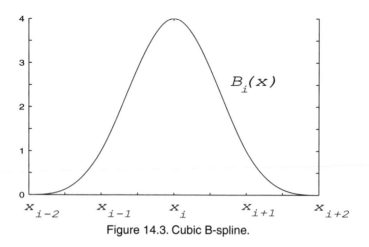

Figure 14.3. Cubic B-spline.

It is understood that translates are set to zero off $[a, b]$. The $n+3$ B-splines $B_i(x)$, $i = -1, \ldots, n+1$, are independent (Exercise 14.7) and hence form a basis for the cubic splines with n equal subdivisions of $[a, b]$. Values of the B-spline (14.12) at knots are tabulated in Table 14.1.

Table 14.1

	x_{i-2}	x_{i-1}	x_i	x_{i+1}	x_{i+2}
$B_i(.)$	0	1	4	1	0
$B_i'(.)$	0	$3/h$	0	$-3/h$	0
$B_i''(.)$	0	$6/h^2$	$-12/h^2$	$6/h^2$	0

RESULT B. There is exactly one cubic spline ψ solving the equally spaced interpolation problem

$$
\begin{aligned}
\psi'(x_0) &= f'(x_0) \\
\psi(x_i) &= f(x_i), \quad 0 \le i \le n \\
\psi'(x_n) &= f'(x_n).
\end{aligned}
\tag{14.13}
$$

PROOF. Following [Prenter], consider the cubic spline

$$
\psi(x) = \sum_{j=-1}^{n+1} c_j B_j(x).
\tag{14.14}
$$

Imposing the interpolation conditions (14.13) yields the system

$$\sum_{j=-1}^{n+1} c_j B_j'(a) = f'(a) \tag{14.15a}$$

$$\sum_{j=-1}^{n+1} c_j B_j(x_i) = f(x_i) \tag{14.15b}$$

$$\sum_{j=-1}^{n+1} c_j B_j(b) = f'(b) \tag{14.15c}$$

of $n+3$ equations in the $n+3$ unknowns c_j. By Table 14.1 this system has matrix

$$A = \begin{pmatrix} -3/h & 0 & 3/h & 0 & 0 & 0 & \cdots & & & \\ 1 & 4 & 1 & 0 & 0 & 0 & \cdots & & & \\ 0 & 1 & 4 & 1 & 0 & 0 & \cdots & & & \\ 0 & 0 & 1 & 4 & 1 & 0 & \cdots & & & \\ & \cdot & & & & & & & & \\ & \cdot & & & & & & & & \\ 0 & 0 & 0 & \cdots & 0 & 1 & 4 & 1 \\ 0 & 0 & 0 & \cdots & 0 & -3/h & 0 & 3/h \end{pmatrix}. \tag{14.16}$$

But A is invertible (Exercise 14.8). Thus because every cubic spline for this partition of $[a, b]$ is of the form (14.14), there must be one and only one cubic spline solving the interpolation (14.13).

Example 3. Let us interpolate $f(x) = \sin x$ on the interval $[0, 2\pi]$ with cubic splines ψ using $n = 4$ equal partitions.

The proof of Result B is actually the algorithm. To obtain the coefficients c_j, invert A and apply its inverse to the column $(f'(a), f(a), f(x_1), \ldots, f(x_i), \ldots, f(b), f'(b))^T$, where in this case $f(x) = \sin x$, $a = 0$, $b = 2\pi$. To graph the resulting spline, save a canonical cubic B-spline with knots at $x = -2, -1, 0, 1, 2$ as a library function, then dilate and translate for the general cubic B-spline. Compare this approach to Exercise 14.9.

But MATLAB has *natural* cubic spline interpolation as a basic command — see Exercise 14.17. Store the function you wish to interpolate (in this case $\sin x$) with

```
function y = fnn(x)
y = sin(x);
```

and then run the following script.

```
% Routine 14.1 Interpolating fnn(x)
a = 0;                            % left endpoint
b = 2*pi;                         % right endpoint
n = 4;                            % number of subintervals
dx = (b - a)/n;                   % increment length
xi = a:dx:b;                      % the knots
yi = fnn(xi);                     % the values at knots
                                  %
x = a:0.01:b;                     % x axis
y = fnn(x);                       % actual function values
                                  %
sp = spline(xi,yi,x);             % spline values at all x
plot(x,y,'--')                    % plot actual values as dashes
hold on;                          %
plot(xi,yi,'o')                   % plot pts.  above knots as ''rings''
plot(x,sp)                        % plot the cubic spline
```

The resulting fit is shown in Figure 14.4 with the interpolated functional values marked on the graph with rings (the symbol o) through which the flexible drafting ruler (spline) must pass.

Example 4. Let us obtain a numerical solution of the retaining wall problem

$$y'''' = xy \qquad (14.17a)$$

subject to the clamped conditions

$$y(0) = y'(0) = 0 = y(1) = y'(1). \qquad (14.17b)$$

After partitioning the interval $[0, 1]$ into n equal subdivisions, our trial solution is to be

$$\tilde{y} = \sum_{j=2}^{n-2} c_j B_j(x). \qquad (14.18)$$

We are justified in discarding the first and last few B-splines since the remaining linear combinations are still dense in $L^2(0, 1)$ by Exercise 14.13. Imposing the Galerkin requirement and moving to the weak restatement, we obtain the system

$$\sum_{j=2}^{n-2} c_j \langle B_j''(x), B_i''(x) \rangle = \langle x B_j(x), B_i(x) \rangle, \qquad (14.19)$$

$i = 2, \dots, n-2$. The solution of this system is left as a major student project (Exercise 14.14).

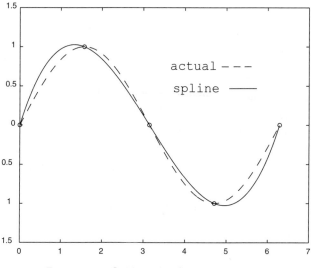

Figure 14.4. Cubic spline fit to $f(x) = \sin x$.

B-splines can be developed in a more unified manner. Following Prenter, I have chosen not to normalize the B-splines of Figures 14.2 and 14.3. But when normalized, the canonical m-B-splines B_i^m at integer knots $x_i = i$ are defined recursively by

$$mB_i^m(x) = (x - i)B_i^{m-1}(x) - (i + 1 + m - x)B_{i+1}^{m-1}(x) \quad (14.20)$$

starting with $B_i^0(x) = 1$ when $i \le x < i+1$, 0 otherwise. They obey recurrences and differential relations much like the theory surrounding orthogonal polynomials (see [Kincaid and Cheney] and Exercises 14.21 through 14.25).

There is significant literature on splines. The best place to start is *A Practical Guide to Splines* by my old classmate Carl de Boor. Also see [Conte and de Boor], [Prenter], and [Schumaker]. The intrepid can attempt [Ciarlet and Lions].

Nowhere is the interest in splines more intense than in Hollywood, where non-uniform rational B-splines (NURBS) are a principal tool of digital animators. Have you ever wanted to be in pictures?

Exercises

14.1 Formally perform the two integration by parts to verify (14.3) using Green's first formula (Exercise 11.16).

14.2 Show that am m-spline with m continuous derivatives is a polynomial of degree m or less.

14.3 Show that it is generically impossible to interpolate both the values $\psi(x_i) = f(x_i)$ and (interior) derivatives $\psi'(x_i) = f'(x_i)$ of a function f with m-splines ψ.

14.4 Find the explicit formula for the basic 2-spline shown in Figure 14.2.

14.5 Find a basis of $S_0^2(\Pi)$. A sketch is sufficient.

14.6 Show that the Hermite interpolation problem

$$\psi(x_i) = f(x_i)$$

$$\psi'(x_i) = f'(x_i)$$

for $0 \le i \le n$ is solvable by a piecewise cubic ψ. Contrast with Exercise 14.3.

14.7 Prove that the $n+3$ translates, $i = -1, \ldots, n+1$, of the cubic B-spline (14.12) are independent over $[a, b]$.

14.8 Prove that the matrix A of (14.16) is invertible.

Outline: Add the first column to the third, then the last to the second to the last. The resulting matrix is diagonally dominant, hence invertible by Gerschgorin's theorem (Exercise 12.13).

14.9 Write a cubic spline interpolation routine to solve (14.13) without recourse to B-splines. Construct the n cubics by solving for their $4n$ coefficients.

14.10 Write a routine to solve the cubic Hermite interpolation of Exercise 14.6.

14.11 Experiment with Routine 14.1. Increase/decrease the number of knots. Fit other functions. Can you find a function f where the fit is bad?

14.12 (Project) Write a quartic spline interpolation routine.

14.13*Show that all linear combinations of the form (14.18) are dense in $L^2(0, 1)$.

14.14 (Project) Complete the retaining wall problem (14.17). Hand calculate the inner products of (14.19), then solve the system for the coefficients c_j. Store the canonical cubic B-spline centered at $x = 0$ with knots $x = -2, -1, 0, 1, 2$ as the function $B(x)$. Using dilations and translations of $B(x)$, graph your solution to (14.17). Compare your solution against the analytic solution in hypergeometric functions given by *Mathematica*.

14.15 (Project) Solve the problem $\nabla^2 \nabla^2 u = 1$ on the unit square $0 < x, y < 1$ subject to clamped boundary conditions via Galerkin finite elements with elements $B_i(x)B_j(y)$. Compare against an analytic solution obtained by separation of variables.

14.16 Partition the entire real line equally with integer knots $x_i = i$. Show that for the (nonnormalized) cubic B-spline of (14.12),

$$f(x) = \sum_{i=-\infty}^{\infty} B_i(x) = 12.$$

Hint: $f'''(x)$ is piecewise constant and periodic modulo 1.

14.17 Prove there is exactly one cubic spline ψ solving the *natural* interpolation problem

$$\begin{aligned} \psi''(x_0) &= 0 \\ \psi(x_i) &= y_i, \quad 0 \le i \le n \\ \psi''(x_n) &= 0. \end{aligned}$$

Why is this the "natural" interpolation?

Hint: Think about bending a flexible ruler.

14.18 Write a routine for performing the natural cubic spline interpolation of Exercise 14.17. Compare your results against MATLAB's `spline`.

14.19 Discuss how to repair continuous but only piecewise differentiable functions f by sniping out small portions of its graph and replacing with cubic splines. Repair $f(x) = |x|$.

14.20 Is it possible to solve the interpolation problem

$$\begin{aligned} \psi(x_0) &= f(x_0) \\ \psi'(x_i) &= f'(x_i), \quad 0 \le i \le n \\ \psi(x_n) &= f(x_n) \end{aligned}$$

with cubic splines?

14.21 Using the recursion (14.20), graph $B_0^m(x)$ for $m = 0, 1, 2, 3, 4, 5$.

14.22 From the recursion (14.20) for the normalized B-splines, prove

$$\sum_{i=-\infty}^{\infty} B_i^m(x) = 1.$$

14.23 Prove that for $m \ge 2$,

$$\frac{d}{dx} B_i^m(x) = B_i^{m-1}(x) - B_{i+1}^{m-1}(x).$$

14.24 Prove

$$\int_{-\infty}^{\infty} B_i^m(x) \, dx = 1.$$

14.25 Prove

$$\int_{-\infty}^{x} B_i^m(y) \, dy = \sum_{j=i}^{\infty} B_j^{m+1}(x).$$

Chapter 15

Technical Writing

Craig Gunn
Technical Writing Consultant
Department of Mechanical Engineering
Michigan State University
East Lansing, Michigan

*You will be judged in industry not on your work
but on how well you communicate your work.*

We dissect the five common instruments of technical communication:
the formal report, the memo, the progress report, the executive summary, and the problem statement. This is followed by a list of *do's*
and *don'ts* for overhead projector presentations. Also included are a
short discussion on how to approach a writing task, several suggestions on style, and a writer's checklist.

§15.1 The Formal Technical Report

The formal report summarizes an analysis, experiment, or simulation
whose results are intended to steer a group's future work toward a
better product for the consumer. The elements of the formal technical report are:

> Title
> Abstract
> Table of Contents
> Nomenclature
> Introduction
> Analysis
> Experimental Equipment and Procedure

Results
Discussion
Conclusions
References
Appendices

Title. The title of the report should be as brief as possible without sacrificing clarity — seven to eight well-chosen words form a typical length. It must provide readers with a clear understanding of the content within.

Abstract (Summary). Although the abstract is placed first, it should not be written until all other parts of the report have been completed. Consisting of only one or two paragraphs (250–300 words), it should state in simple declarative sentences what was attempted, how it was accomplished (but only if special techniques were utilized), and what was achieved. That is, it should highlight the main points of the report.

Although this is the shortest and most difficult section to write, it is the most important. In technical publications, the abstract, and only the abstract, is what most people will read. Therefore it *must stand alone*, presenting all relevant ideas and results to those readers who will not go into the report for explanation. Often it is printed separately from the report.

Table of Contents. The table of contents lists each section heading along with the page where it can be found in the report. The table of contents itself is not listed, nor is the abstract. A list of illustrations and a list of tables will follow the table of contents if there are figures and tables in the report.

Nomenclature. The nomenclature list defines all symbols used in the report alphabetically: first Latin, then Greek. Readers need to have a place where they can easily find an explanation of the symbols used in the analysis section. When acronyms and abbreviations are used extensively, they should not be placed with the nomenclature, but rather should be placed in a separate listing. Acronyms and abbreviations should always be spelled out at their first appearance in the text, e.g., fast Fourier transform (FFT).

Introduction. The introduction should describe the motivation for the project and provide any germane background information. Its essential purpose is to present the substance of the work and the context in which it was executed. Discuss here why you have chosen to perform this task — be certain that you refer to the product that is being investigated and the importance of your work to its successful development. At the end of the introduction, briefly describe the material contained in the report by noting what is presented in each of the sections to follow. The introduction provides the motivation to read on.

Analysis. The underlying mathematical basis of the project is presented in this section. The analysis should proceed from the general (and well-known) basic relationships to the specific formulas used in the interpretation of the data. The symbols must be clearly defined or should at least appear in the nomenclature list described earlier.

Analytical results that have been derived previously and are readily available (e.g., equations from a published text) can be quoted with suitable reference. Their derivation need not be repeated as long as that derivation is not central to an understanding of the work.

All relevant mathematical analysis should be presented. What is crucial to a good analysis are the supporting explanations and commentary on the mathematics. Do not require the reader to consult laboratory handouts or textbooks to understand the specific analysis required for the experiment.

Experimental Equipment and Procedure. A schematic representation of the experimental equipment, including detailed views of unusual or important components, is presented and discussed in this section. Other illustrations can be used to document pertinent dimensions of the apparatus and to further specify details of the experimental equipment used for the study. Tables, which are typically constructed to present information in numerical form (e.g., percentages, dimensions, and mechanical and electrical data), are also extremely helpful in this section.

If the procedure used for the experiments is not an established one, details of the techniques must be included. The criterion is that someone familiar with the general area of investigation should be able to reproduce your experiments from the information provided in this section.

Remember that the report's flow and overall feel can be destroyed

by an incomplete, or inaccurate discussion of the experimental procedure and equipment or by graphical or tabular matter that is mis-
placed, unexplained, incorrectly captioned, or inappropriately sized.
Regarding the latter, some major points to remember are:

- When first referencing a figure or a table, place it on the same
  page or page after its explanation. It is distracting for the
  reader to have to leaf through the report to find a clarifying
  figure/table.

- Make figures at least one-third of a page in size with call-outs in
  8- to 10-point bold type. Small figures and callouts are difficult
  to read.

- Include clear, complete, and sequentially numbered captions
  for each figure and table.

- Provide readers with a clear, precise explanation of each figure
  and table.

- Never include figures or tables that are not mentioned in the
  text!

The guidelines above apply to figures and tables used throughout
the text.

Results. This is the section where the answers obtained from the
analysis and experiments are presented. It should contain concise
statements, referring often to the graphic or tabular data of the report. These statements inform the reader (without discussion) of the
author's interpretation of the results, based upon the analysis and
experimental data.

Never forget that your readers may have picked up your report
in midstream — they may have read the abstract and then gone
directly to the results. Always assume that the report will not be
read from the first page to the last. With this in mind, make the
results section count.

Discussion. The discussion begins with a very brief summary of the
results and then proceeds to their interpretation by noting what is
"as expected," what is unexpected, and what is of primary technical
interest. This interpretation of the results, in terms of the motivation

for the experiment and its relevance to current projects where you work or study, should be the focus of this section.

The discussion could involve a comparison with other similar investigations or a comparison with expected results. Experimental results should be described precisely, to the accuracy of the measurements, and the estimated uncertainties and their effect on the calculated values should be noted. It is extremely important to provide exact statements and avoid vague modifiers such as *greater than, about, like,* and *sort of.* Any limitations of the work should be emphasized here along with the strong points — if you do not point out the limitations, someone else surely will later.

Be aware of how you are presenting information to readers. Readers have specific expectations — they expect to receive information in a pattern that presents what is known first and then what is new second. Begin with old or known information and then introduce new, related information. In other words, the discussion must follow a logical progression in order to soundly support the conclusions of the next section.

Conclusions. A useful beginning is to state that the following conclusions are supported by the results of this study. Then list these conclusions in one or more simple (declarative) sentences, with bullets or numbers to separate each one. Remember that readers are looking for concise statements of your results and discussion. They are not interested at that point in further explanation. They want the masses of data tidily synthesized into a brief set of conclusions. Never introduce new material in this section.

References. A list of cited references appears in this section, with great care taken to follow the specified format of the organization for whom the report is being written. A sample format might cite references by number sequentially in the text, e.g., [1], [2], ..., and then list them in the reference section at the end of the report in the order cited.

[1] R. E. Walker, A. R. Stone, and M. Shandor, "Secondary gas injection in a conical rocket nozzle," *AIAA Journal,* vol. 1, no. 2, February 1963, pp. 334–338.

[2] M. J. Turner, H. C. Martin, and R. C. Leible, "Further development and applications of stiffness method," *Matrix Methods*

of Structural Analysis, 2nd ed., vol. 1, Macmillan, New York, 1964, pp. 203–206.

[3] E. Segré, ed., *Experimental Nuclear Physics,* vol 1, Wiley, New York, 1953, pp. 6–10.

[4] E. Book and H. Bratman, *Using Compilers to Build Compilers,* SP-176, Systems Development Corporation, Santa Monica, CA, August 1960.

[5] S. L. Soo, "Boundary layer motion of a gas–solid suspension," *Proceedings of the Symposium on Interaction Between Fluids and Particles,* Institute of Chemical Engineers, vol. 1, 1962, pp. 50–63.

Always provide page numbers for journal articles and a page or chapter number for books. Never list references that are not cited in the text.

Appendices. Lengthy calculations or side issues that would interrupt the main thrust of the report should be relegated to an appendix. A primary criterion for deciding whether or not to put something in the main body of the report is to ask if its inclusion is required for the logical evolution of the ideas? If not, the item should either be left out or placed in an appendix.

Bear in mind that these are only the briefest comments about a very important tool of engineering and science. It is likely that you will spend a great deal of your time in the preparation of technical reports. Your skill in this area will in large measure determine how far you advance in your profession.

EXERCISE 15.1. Using the format above , prepare a formal report on a topic of your choice or on one assigned by your instructor.

§15.2 The Memo

Memos are written by all "knowledge workers." They are most often designed to be used within the writer's own company or organization. They may note the existence of a problem, propose some course of action, describe a procedure, or report the results of a test or an investigation. They are sometimes referred to as informal writing, which is not the same as sloppy, casual, or careless writing. A memo must be carefully prepared, thoughtfully written, and thoroughly proofread for errors. It begins with headings:

To: Name, Job Title
         Department
         Name of Organization

From: Name, Job Title
         Department
         Name of Organization

Subj: Issues addressed in the memo

Date: Date

         with optional headings:

Dist: Distribution list of other people receiving the memo

Encl: Other documents included with the memo

Ref: List of important background documents

The format of the memo is simple and contains the following parts:

- Foreword — presents the statement of the problem or important issue.

- Summary — may include an approach to resolving the problem or issue, the details of a procedure or investigation, or any other information that a reader needs to know.

- Discussion or Details — provides the specialized information needed by technically involved readers and support for any claims in the summary.

EXERCISE 15.2. Prepare a memo to your instructor explaining a current project or assignment that you are working on, a problem for which you have a solution, or an idea that you would like to propose.

§15.3 The Progress Report

As you move from task to task you will be required to keep your managers and co-workers informed of your activities. The progress report is used to clearly and concisely summarize your efforts on a particular project. Much like any other report prepared for an interested party, it will consist of sections:

Beginning. This gives an indication of what the report is about and where it is going to lead. Expose the readers to the scope and objective of the task at hand, the overall progress to date, and any changes to the initial goals.

Middle. Here the reader is acquainted with the specific time span of the project and exactly what progress has been achieved so far. Describe in detail the work that has been done, the tasks that need to be completed, and the schedule that has been set for their timely accomplishment.

End. In this section you pull all the information in your report into a logical, coherent whole. Summarize the main points and reiterate where you are along the road to completion.

Remember that this report may be used to stop the effort in midstream or to assign help for its realization. For the sake of the project, then, make certain that the report is attractive and easy to understand.

EXERCISE 15.3. If you are involved in a project for one of your courses, prepare a progress report to inform your colleagues of its status.

§15.4 The Executive Summary

Often you may be required to submit an executive summary rather than a full report. The executive summary condenses all the work done on a project into the briefest of documents. It will contain information that a nontechnically oriented person will need to make a clear decision on the work that you have done. Pare the content to only what has been accomplished, how it has been accomplished, and at what cost. Describe the key elements of your work concisely in language that will be understood by a reader who may only vaguely understand the technical basis of the project. The executive summary focuses on the managerial side of business, not on the technical side. It should contain the following information:

- The background of the situation or the problem

- How the work has been accomplished

- Implications of cost

- Conclusions

- Recommendations

Observe the similarities between an executive summary and an abstract. Both communications are short in length. Each relies upon concise, clear language to focus the reader's attention on what is being investigated, what tasks have been accomplished, what conclusions have been drawn, and what course of action should be taken. In many cases, this will be the only text that is read by management before a decision to proceed is made. With that understanding, you will need to expend much careful effort in preparing both of these communications.

EXERCISE 15.4. From a report that you have already prepared, write an executive summary to be presented to upper management.

§15.5 The Problem Statement

During the course of your work on engineering and science projects, you may be asked to present a problem statement concerning the work that you assume is to be done and the course that you will take to complete it. It is important that the initial statement of the problem be as clearly communicated as the solution itself. Here are some basic guidelines to follow before you begin to write:

- Working with data that you have collected, pursue all avenues that might lead to a solution.

- Use personal contacts with colleagues and others who can share their knowledge of the problem with you.

- Actively investigate the problem firsthand.

- Make sure that all your information is accurate and complete.

Once the data are collected, you can begin to prepare the problem statement. It will contain a direct statement of what you perceive the particular problem to be. Do not address peripheral issues. A clear focus on the true problem will assist you immeasurably in logically presenting the approach to its solution. The document should fully describe each step that you will take to bring a successful conclusion to the work.

EXERCISE 15.5. Select a problem of interest to you. Evaluate what you believe to be the true basis of the problem and how you will go about achieving a solution.

§15.6 Overhead Projector Presentations

Many researchers say that your skill in oral presentation is more crucial for success than your writing. Let us sketch the *do's* and *don'ts* of that fond institution of the technical world, the overhead projector talk. In preparing your slides:

- Do not "slap and snatch" — plan at least 1 minute per slide.

- Do not dwell more than 3 minutes on one slide.

- Never use more than seven bullets per slide.

- Lead the listener logically down your chosen path.

- Do not present detail — distill each idea into one short phrase.[1]

- Provide "hooks" to bring distracted listeners back into the flow.

- Use overlays.

- Be cautious with color.

- Use readable type.

Some believe that the title slide must be followed by a talk outline slide and a problem definition slide. But all agree with the following rule:

---

[1]But not a cryptic phrase. As a rule of thumb, a listener should be able to reconstruct an overview of your presentation given copies of your slides.

Tell them what you are about to tell them, tell them, then tell them what you have told them.

With the technology available today, there is no excuse for shoddy handwritten slides. Slides should be prepared via computer software and then laser printed in at least 24-point type. Be cautious with colored backgrounds which may wash out text. Test color combinations on the projector before giving the talk.

§15.7 Approaching a Writing Task

As you approach a writing task, study your audience to determine how they will receive your communication. For instance, by knowing your audience, you may be able to defuse conflict about a controversial topic and bring your readers into a positive frame of mind. Think about the way that your audience perceive the world around them and tailor your communication accordingly.

Once the nature of your audience has been determined, the next step is to construct a basic outline to ensure the orderly flow of your material. Then, thanks to electronic word processing, the initial writing process becomes a brainstorming session:

- Write without hesitation or concern for quality.

- Put ideas on paper as quickly as possible.

- Do not revise as you go.

This first burst of writing further clarifies the problem — it reveals what is known and what is not, provides a basis upon which to seek new information, clearly indicates gaps in information, and gives a foundation upon which to build. After this initial frenzy, the hard work of revision begins.

§15.8 Style

Use appropriate language. Because you want your reader to understand your message, it is important to fit the language to the audience. Since most of your text will be directed at audiences comprised of scientists and engineers, you should learn to avoid language that does not suit a technical writer. Technical writing is direct, concise, and clear. If you write to impress, you will fail many more times than you succeed.

It is a careless writer who uses slang, cliche's, ornate words, and meaningless jargon. In other words, if you wish to keep your reader's attention, do not dress up ideas in tasteless, trite, stilted, or official-sounding expressions. Rather, use simple, natural language that will clearly express what you mean.

It is extremely important in a world where jobs are performed by both males and females to avoid sexist language. Words that designate gender distinctions must be removed. For example, a pronoun such as *he* used to described a mixed group creates a biased view. One of the easiest ways to eliminate gender focus is to make the words plural, thereby removing the male or female focus. Another way is to replace words that unnecessarily distinguish between male and female, e.g., *employee* for *workman* and *chairperson* for *chairman*.

Stay away from odd-looking and sounding mutations that do not help the text to flow, such as *s/he, he/she,* or *his/or/her.* Use plurals, change words, or simply say *he and she, his or her,* or *him and her.*

Enhance flow. In a text that is fluid, the reader will sense that the ideas are moving smoothly toward some conclusion. How is this achieved?

One way of enhancing flow is to resurrect the *pronoun.* It is very appropriate to use pronouns to draw needed connections between text. "The lab was a mixture of difficult equations and painstaking work. It was not meant to be an easy exercise." The reader is drawn into the second sentence and sees the connection through the use of *it,* which refers back to the lab experiment. These pronouns serve as a link between sentences currently being read and those already finished.

Repetition of keywords (for emphasis) can also help your readers. However, remember that although repetition helps to reinforce your ideas, needless overuse of a word will only make for tedious reading. Always carefully evaluate each word that you use to provide the reader with effective continuity of thought.

Transitional devices allow the reader to see connections between varying ideas and statements. The lack of transition is a primary cause for reader disinterest and frustration. Transitions in all their forms allow the reader to move easily through a text — without these simple signposts, a document can become lifeless and difficult to read. One example of this device is the *transitional word.* Such words help a reader navigate through text by connecting one idea to the next, one sentence to another, and one paragraph to another, e.g., *first, then, on the other hand, besides, furthermore, therefore, similarly, in lieu of this, likewise, finally, as a result, however,* and *moreover.*

Another approach for enhancing the flow of ideas involves *varying sentence and paragraph length.* This technique allows the reader to enjoy the "experience of reading" what is often highly complex, monotonous technical text. Making a group of colorless sentences more fluid and readable involves creating structures that vary in length from the simplest of sentences (subject/verb) to those that are extremely complex (two or more main clauses and one or more subordinate clauses). In terms of paragraph length, very short paragraphs should be avoided because they do not allow the reader to see groupings of sentences that are related. On the other hand, overly lengthy paragraphs obliterate subdivisions in thought — there is nothing more disruptive to the reader than a series of paragraphs that consistently run to more than one page.

When we read or listen, we like to sense *continuity.* We enjoy being able to follow the train of ideas, i.e., how one idea leads to another. It is disconcerting to be given one piece of information only to be jolted by another unrelated piece, with no hint as to its connection. It is vitally important that you be aware of how each sentence combines with its neighbors to create a flowing, coherent whole.

Within the paragraph, the main focus usually rests in the *topic sentence.* The topic sentence can appear at any point in the paragraph. It also can appear in the form of a pervading idea in which the reader is clearly able to see the focus of the paragraph. It is your job to make certain that all the ideas contained within a paragraph relate to one central theme. New ideas require new paragraphs.

Be consistent. If you capitalize Figure and Table in the beginning of the text, continue throughout the text. If you capitalize something on one page, do not lower its case on the next page. Make certain that it is used appropriately the first time and then continue to use it correctly as you progress through the text. Your readers may not be aware of the first instance of your indecisiveness, but eventually they will notice the vacillation.

Proofread your text. There is nothing more horrifying than reading a document that is full of spelling and grammatical mistakes. A spell checker cannot think; it does not know the word you wanted to use, e.g., it cannot decide that you meant to use *form* (not *from*) or *trial* (not *trail*). This cannot be emphasized enough — carefully proofread your text!

Read what you have written out loud. Where you stumble, there is a problem. Mark those places that cause you difficulty and make an effort to improve their flow. Sometimes it is easier to write a new paragraph or even a subsection rather than doctor a faulty one.

EXERCISE 15.8A. Evaluate the number of transitional words that are present in a piece of text. Look carefully at how these words make the sentences flow together. If the text does not contain any visible transitional words, insert your own to improve the reader's understanding.

EXERCISE 15.8B. Read two pages of technical text. Carefully analyze the sentence structure. Is there variety of structure, and does it help you to read the text?

EXERCISE 15.8C. Select four different pieces of text: a children's book, a newspaper, a legal document, and a technical article. Use the points of style above to evaluate each piece of text.

Yearly reread W. Strunk and E. B. White's *The Elements of Style,* 3rd ed. (Macmillan, New York, 1979).

§15.9 Writer's Checklist

- Is the purpose of the document clear?

- Are each of the headings in the outline covered?

- Does the text flow logically?

  - Do the sections and paragraphs have a central focus?

  - Are stylistic techniques such as the use of transitional words and repetition employed?

- Are there contradictions in the document?

- Are there indefinite or vague statements?

- Does the information in the document prepare the reader for the conclusions?

- Do the data support the conclusions?

- Are the conclusions clear and logical?

- Will readers draw the same conclusions as the author does? Or are there portions of the text that could be interpreted in a variety of ways?

- Are the recommendations appropriate?

- Has the text been proofread?

- Is the document written so that the nontechnical reader can understand it?

- Is the writing clear enough and persuasive enough to be accepted by the most skeptical readers?

In the end, remember that technical writing is truthful, is disinterested, is logically developed, contains no emotion, contains no unsupported opinions, is sincere, is not argumentative, and does not exaggerate.

References

A. B. Abel and B. S. Bernayke, *Macroeconomics*, 3rd ed., Addison-Wesley, Reading, MA, 1998.

L. V. Ahlfors, *Complex Analysis*, McGraw-Hill, New York, 1966.

J. Baldani, J. Bradfield, and R. Turner, *Mathematical Economics*, Dryden Press, Fort Worth, TX, 1996.

B. Batchelor and F. Waltz, *Interactive Image Processing for Machine Vision*, Springer-Verlag, London, 1993.

A. Beiser, *Applied Physics*, 2nd ed., Schaum's Outline Series, McGraw-Hill, New York, 1988.

W. Bialek, "Thermal and quantum noise in the inner ear," *Mechanics of Hearing*, deBoer and Viergever, ed., Martinus Nijhoff Publishers, The Hague, The Netherlands, 1983, pp. 185–192.

W. E. Boyce and R. C. DiPrima, *Elementary Differential Equations and Boundary Value Problems*, 6th ed., Wiley, New York, 1996.

W. L. Briggs and V. E. Henson, *The DFT*, SIAM Publications, Philadelphia, 1995.

W. C. Brown, *Matrices over Commutative Rings*, Marcel Dekker, New York, 1993.

J. A. Cadzow, *Discrete-Time Systems*, Prentice Hall, Englewood Cliffs, NJ, 1973.

R. V. Churchill and J. W. Brown, *Fourier Series and Boundary Value Problems*, 4th ed., McGraw-Hill, New York, 1987.

P. G. Ciarlet and J. L. Lions, ed., *Handbook of Numerical Analysis*, North-Holland, Amsterdam, 1991.

R. R. Clements, "Reliability of multichannel systems," *Mathematical Modelling: Classroom Notes in Applied Mathematics*, M. S. Klamkin, ed., SIAM Publications., Philadelphia, 1987, pp. 256–264.

C. W. Cobb and P. H. Douglas, "A theory of production," *American Economic Revue Supplement*, vol. 18, 1928, pp. 139–165.

S. D. Conte and C. de Boor, *Elementary Numerical Analysis,* 2nd ed., McGraw-Hill, New York, 1972.

R. Courant, *Calculus of Variations,* Courant Institute, New York University Press, New York, 1962.

R. Courant and D. Hilbert, *Methods of Mathematical Physics,* vols. 1 and 2, Wiley, New York, 1989.

H. Cramér, *Mathematical Methods of Statistics,* Princeton University Press, Princeton, NJ, 1961.

C. Cussler and C. Dirgo, *The Sea Hunters,* Pocket Books (Simon and Schuster), New York, 1997.

C. de Boor, *A Practical Guide to Splines,* Springer-Verlag, New York, 1978.

C. Derman and S. M. Ross, *Statistical Aspects of Quality Control,* Academic Press, San Diego, CA, 1996.

J. J. DiStefano, A. R. Stubberud, and I. J. Williams, *Feedback and Control Systems,* 2nd ed., Schaum's Outline Series, McGraw-Hill, New York, 1990.

R. C. Dorf, *Modern Control Systems,* 4th ed., Addison-Wesley, Reading, MA, 1986.

R. Dorfman, P. A. Samuelson, and R. M. Solow, *Linear Programming and Economic Analysis,* Dover, New York, 1987.

P. DuChateau and D. W. Zachmann, *Theory and Problems of Partial Differential Equations,* Schaum's Outline Series, McGraw-Hill, New York, 1986.

J. A. Duffie and W. A. Beckman, *Solar Engineering of Thermal Processes,* Wiley, New York, 1980.

J. A. Edminister, *Electromagnetics,* 2nd ed., Schaum's Outline Series, McGraw-Hill, New York, 1993.

Energy Information Administration, *Annual Energy Outlook, with Projections to 2010,* DOE/EIA-0383(91), 1991.

EPRI, *End-Use Technical Assessment Guide: Fundamentals and Methods,* EPRI CU-7222 V4, Electric Power Research Institute, Palo Alto, CA, 1991.

A. Erdelyi, *Operational Calculus and Generalized Functions,* Holt, Rinehart and Winston, New York, 1962.

V. Fabian and J. Hannan, *Introduction to Probability and Mathematical Statistics,* Wiley, New York, 1985.

R. P. Feynman, *Statistical Mechanics,* Lecture Note Series, Addison-Wesley, Redwood City, CA, 1990.

J. B. J. Fourier, *Théorie Analytique de la Chaleur,* 1822.

E. G. Frankel, *Systems Reliability and Risk Analysis,* 2nd ed., Kluwer Academic Press, Boston, 1988.

J. Franklin, *Methods of Mathematical Economics,* Springer-Verlag, New York, 1980.

M. Frazier, *An Introduction to Wavelets through Linear Algebra,* Springer-Verlag, New York, 1999.

A. Friedman and W. Littman, *Industrial Mathematics,* SIAM Publications, Philadelphia, 1994.

F. M. Gardner, *Phaselock Techniques,* 2nd ed., Wiley, New York, 1979.

N. H. Gardner, C. Messer, and A. K. Rathi, ed., *Monograph on Traffic Flow Theory,* report prepared for the Federal Highway Administration, 1996.

D. L. Gerlough and M. J. Huber, *Traffic Flow Theory,* Special Report 165, Transportation Research Board, National Research Council, Washington, DC, 1975.

F. R. Giordano, M. D. Weir, and W.P. Fox, *A First Course in Mathematical Modeling,* 2nd ed., Brooks/Cole, Pacific Grove, CA, 1997.

H. Goldstein, *Classical Mechanics,* 2nd ed., Addison-Wesley, Reading, MA, 1981.

G. H. Golub and J. M. Ortega, *Scientific Computing and Differential Equations,* Academic Press, San Diego, CA, 1992.

M. Hall, *Combinatorial Theory,* 2nd ed., Wiley, New York, 1986.

P. R. Halmos, *An Introduction to Hilbert Space,* 2nd ed., Chelsea, New York, 1957.

W. M. Hartmann, *Signals, Sound, and Sensation,* American Institute of Physics Press, Williston, VT, 1997.

F. S. Hillier, *Operations Research,* 2nd ed., Holden-Day, San Francisco, 1974.

F. C. Hoppensteadt, *Mathematical Methods of Population Biology,* Cambridge University Press, New York, 1982.

R. Jain, R. Kasturi, and B. Schunck, *Machine Vision,* McGraw-Hill, New York, 1995.

N. Keyfitz, *Applied Mathematical Demography,* 2nd. ed., Springer-Verlag, New York, 1977.

N. Keyfitz and W. Flieger, *Population: Facts and Methods of Demography,* W.H. Freeman, San Francisco, 1971.

D. Kincaid and W. Cheney, *Numerical Analysis,* 2nd ed., Brooks/Cole, Pacific Grove, CA, 1996.

M. S. Klamkin, *Mathematical Modelling: Classroom Notes in Applied Mathematics,* SIAM Publications, Philadelphia, 1987.

M. W. Klein, *Mathematical Methods for Economics,* Addison-Wesley, Reading, MA, 1998.

H. Kolsky, *Stress Waves in Solids,* Dover, New York, 1963.

J. D. Kraus, *Antennas,* 2nd ed., McGraw-Hill, New York, 1988.

O. A. Ladyzhenskaya, *The Boundary Value Problems of Mathematical Physics,* Springer-Verlag, New York, 1985.

D. Lancaster, *Active Filter Cookbook,* H.W. Sams & Co., Indianapolis, IN, 1975.

R. Layard and S. Glaister, ed., *Cost-Benefit Analysis,* 2nd ed., Cambridge University Press, Cambridge, 1996.

C. R. MacCluer, "Stability from an operational viewpoint," *IEEE Transactions on Automatic Control,* vol. 33, no. 5, May 1988.

C. R. MacCluer, *Boundary Value Problems and Orthogonal Expansions,* IEEE Press, Piscataway, NJ, 1994.

A. D. May, *Traffic Flow Fundamentals,* Prentice Hall, Englewood Cliffs, NJ, 1990.

P. J. McCleer, C. R. MacCluer, and S. E. Zilinski, *Potential Use and*

Benefits of Advanced Components in a 1993 Time Frame Commercial Unitary Heat Pump, report sponsored by the Electric Power Research Institute, Palo Alto, CA, 1992.

R. McOwen, *Partial Differential Equations,* Prentice Hall, Upper Saddle River, NJ, 1996.

J. Mikusiński, *Operational Calculus,* 2nd ed., vol. I, Pergamon Press, Elmsford, NY, 1983.

S. K. Mitra and J. F. Kaiser, ed., *Handbook for Signal Processing,* Wiley, New York, 1993.

P. M. Morse and G. E. Kimball, *Methods of Operations Research,* MIT Press and Wiley, New York, 1951.

P. J. Nahin, *Oliver Heaviside: Sage in Solitude,* IEEE Press, New York, 1987.

P. V. O'Neill, *Advanced Engineering Mathematics,* 2nd ed., Wadsworth, Belmont, CA, 1987.

W. J. Orvis, *1-2-3 for Scientists and Engineers,* Sybex, San Francisco, 1987.

A. Papoulis, *Signal Analysis,* McGraw-Hill, New York, 1977.

D. A. Pierre, *Optimization Theory with Applications,* Dover, New York, 1986.

G. Polya, *How to Solve It,* 2nd ed., Princeton University Press, Princeton, NJ, 1985.

P. M. Prenter, *Splines and Variational Methods,* Wiley, New York, 1975.

T. Puu, *Nonlinear Economic Dynamics,* 4th ed., Springer-Verlag, Berlin, 1997.

L. Råde and B. Westergren, *Beta Mathematics Handbook,* 2nd ed., CRC Press, Boca Raton, FL, 1990.

F. Riesz and B. Sz-Nagy, *Functional Analysis,* Ungar, New York, 1955.

C. B. Rorabaugh, *Digital Filter Designer's Handbook,* McGraw-Hill, New York, 1997.

W. Rudin, *Real and Complex Analysis,* 3rd ed., McGraw-Hill, New York, 1987.

T. P. Ryan, *Statistical Methods for Quality Improvement*, Wiley, New York, 1989.

D. Salvatore, *Microeconomic Theory*, 3rd ed., Schaum's Outline Series, McGraw-Hill, New York, 1992.

L. L. Schumaker, *Spline Functions: Basic Theory*, Wiley, New York, 1981.

I.M. Sobol, *A Primer for the Monte Carlo Method*, CRC Press, Boca Raton, FL, 1994.

S. L. Sobolev, *Partial Differential Equations of Mathematical Physics*, Dover, New York, 1964.

E. D. Sontag, *Mathematical Control Theory*, Springer-Verlag, New York, 1990.

M. R. Spiegel, *Statistics*, 2nd ed., Schaum's Outline Series, McGraw-Hill, New York, 1996.

H. M. Steiner, *Engineering Economic Principles*, rev. ed., McGraw-Hill, New York, 1992.

G. Strang, *Introduction to Applied Mathematics*, Wellesley-Cambridge Press, Wellesley, MA, 1961.

G. Strang and G. J. Fix, *An Analysis of the Finite Element Method*, Prentice Hall, Upper Saddle River, NJ, 1973.

T. Svobodny, *Mathematical Modeling for Industry and Engineering*, Prentice Hall, Upper Saddle River, NJ, 1998.

G. Taguchi and Y. Wu, *Introduction to Off-Line Quality Control*, Central Japan Quality Control Association, 1979 (available from American Supplier Institute, 32100 Detroit Industrial Expressway, Romulus, MI 48174).

H. M. Wagner, *Principles of Operational Research, with Applications to Managerial Decisions*, Prentice Hall, Englewood Cliffs, NJ, 1969.

M. H. Wright, "A brief history of linear programming — from Fourier to the ellipsoid algorithm," *SIAM News*, March 1989, pp. 10–11.

N. Young, *An Introduction to Hilbert Space*, Cambridge University Press, Cambridge, 1988.

Index